Strategic Planning, Execution, and Measurement (SPEM)

A POWERFUL TOOL FOR CEOs

Strategic Planning, Execution, and Measurement (SPEM)

A POWERFUL TOOL FOR CEOs

Girish P. Jakhotiya

CRC Press
Taylor & Francis Group
Boca Raton London New York

CRC Press is an imprint of the
Taylor & Francis Group, an **informa** business

A PRODUCTIVITY PRESS BOOK

CRC Press
Taylor & Francis Group
6000 Broken Sound Parkway NW, Suite 300
Boca Raton, FL 33487-2742

© 2013 by Girish P. Jakhotiya
CRC Press is an imprint of Taylor & Francis Group, an Informa business

No claim to original U.S. Government works

Printed on acid-free paper
Version Date: 20130401

International Standard Book Number-13: 978-1-4665-6745-0 (Paperback)

This book contains information obtained from authentic and highly regarded sources. Reasonable efforts have been made to publish reliable data and information, but the author and publisher cannot assume responsibility for the validity of all materials or the consequences of their use. The authors and publishers have attempted to trace the copyright holders of all material reproduced in this publication and apologize to copyright holders if permission to publish in this form has not been obtained. If any copyright material has not been acknowledged please write and let us know so we may rectify in any future reprint.

Except as permitted under U.S. Copyright Law, no part of this book may be reprinted, reproduced, transmitted, or utilized in any form by any electronic, mechanical, or other means, now known or hereafter invented, including photocopying, microfilming, and recording, or in any information storage or retrieval system, without written permission from the publishers.

For permission to photocopy or use material electronically from this work, please access www.copyright.com (http://www.copyright.com/) or contact the Copyright Clearance Center, Inc. (CCC), 222 Rosewood Drive, Danvers, MA 01923, 978-750-8400. CCC is a not-for-profit organization that provides licenses and registration for a variety of users. For organizations that have been granted a photocopy license by the CCC, a separate system of payment has been arranged.

Trademark Notice: Product or corporate names may be trademarks or registered trademarks, and are used only for identification and explanation without intent to infringe.

Visit the Taylor & Francis Web site at
http://www.taylorandfrancis.com

and the CRC Press Web site at
http://www.crcpress.com

Dedicated to my clients!

Contents

Preface xi
Author xiii

1 Strategic Planning: A Tool for Reaching World-Class Status 1
  Defining World-Class Status 1
  Gaining World-Class Status 2
  Benefits of Strategic Planning 6
  Fundamental Factors Defining Scope of Strategic Planning 7
    Institutionalization of Vision and Mission 8
    Example 1.1: Indian Automobile Company 10
    Other Important Factors: Business Model, Core Competence,
    and Value Chain 11
      Business Model and Return on Investment 12
      Core Competence 14
      Value Chain 15
    Example 1.2: A Company and B Company 16
    Example 1.3: ABC Company 18

**2 Prerequisites of Strategic Planning:
  Inquiry and Status Evaluation** 21
  Prerequisite 1: Examining Performances, Perspectives,
  and Aspirations 21
    Part A: General Inquiry 22
    Part B: Business Development Inquiry 34
    Part C: Strategic Cost Management Inquiry 40
    Part D: Operations Management Inquiry 45
    Part E: Employee Productivity Inquiry 50
    Part F: Resource Management Inquiry 60
  Prerequisite 2: Status Evaluation 66
    Versatile Corporate Evaluation 67

3 Strategic Plan: Components, Preparation, Review, and Recommendations 73
Structure and Components 73
Duration 74
Format 74
    Strategy for Performance Areas 74
    Hurdle Management 75
    Cost–Benefit Analysis (CBA) 75
    Schedule for Execution 75
    Crisis Management and Exit Route 75
Execution and Impact Routes 76
    Execution Route 76
    Impact Route 76
Preparation 76
Implementation 79
Review and Recommendations for Improvisation 80
Generic Strategies for Five Performance Areas 82

4 Monitoring Parameters for Strategy Execution and Impact Measurement 119
Need for Monitoring 119
Periodicity of Monitoring 120
Monitoring Process 120
Monitoring Parameters 122
    Business Growth and Expansion Monitoring Tools 124
    Strategic Cost Management Monitoring Tools 127
    Operations, Logistics, and Technology Management Monitoring Tools 131
    Employee Development and Satisfaction Monitoring Tools 134
    Resource Management Monitoring Tools 140

5 Measurement of Strategy Execution and Financial Monitoring 145
Pyramid Movement of ROI for Strategy Execution 145
Financial Impacts of Strategy Execution 146
    Financial Monitoring 149
        Common Characteristics of Financial Parameter Tables 152
        Level 1: Systemic Employees 155

       Level 2: Knowledge Employees 174
       Level 3: Decision-Making Employees 186
       Level 4: Entrepreneurial Employees... 192

6 Entrepreneurial Score Card: A Tool for Improving Impacts of Strategic Planning and Execution 197

Need for Entrepreneurial Score Card (ESC) 197
Ten Characteristics of Entrepreneurial Employees.............................. 199
Entrepreneurial Behavior Questionnaire..200
Classification of Employees Based on Competencies, Roles, and Empowerment Levels 201
       ESCs for All Employees..202
ESCs for Four Employee Categories from Six Functions 203
       Sales and Marketing Employees...208
       Production or Operations Employees... 212
       Purchasing Employees 216
       Finance and Accounts Employees 220
       Human Resource (HR) Employees 227
       Research and Development Employees..233
Entrepreneurial Rewards..239
Knowledge Management for Entrepreneurial Competencies................ 241
       Objectives 241
       Key Components 242
           Methods of Identifying Knowledge Requirements................242
           Knowledge Sources 242
           Knowledge Classification and Storage 242
           Knowledge Evaluation for Potential Use 242
           Knowledge Sharing 243
           Knowledge Application 244
           Knowledge Sharing Mechanisms 244
           Employee Feedback..244
           Management Appraisal 246
       Promotion of Knowledge Management Efforts...........................246

7 Comprehensive Case Study 249

Background and Recent Performance..249
       SWOT Analysis 252
Strategic Plan: Strategies for Five Performance Areas...........................252

Measuring Financial Impact of Plan on Company Performance............263
    Budgeted Performance and Result Indicators for
    the Year 2013–2014..265
Execution Route, Time and Resource Mapping, and Impact Route......268

Index 271

Preface

During the past two decades, the global economy has gone through many unpredicted episodes of turbulence. Uncertainties give birth to challenges as well as to opportunities. Many small- and medium-size business enterprises and their ambitious promoters successfully exploited these opportunities and grew very rapidly. But this growth was not the result of proper strategic planning. Rather, it was due to successful short-term decisions. For sustainable growth, a strategic plan is a must.

In cases of large business organizations, young and brilliant executives are looking for fast entrepreneurial growth. So they have no choice but to think strategically and achieve sustainable growth. It is the modern trend to involve young executives in the critical and comprehensive process of strategic planning.

Quite a few CEOs possess entrepreneurial vision and have the operational excellence to execute relevant missions. However, many lack the skill to articulate the painful process of preparing a robust and reasonable strategic plan. They are compelled to depend on consultants to develop appropriate methodologies to prepare relevant strategic plans. CEOs and their colleagues should acquaint themselves with these methodologies so they can use consultants' services to add better value rather than plan corporate strategies.

Keeping in mind all the above requirements, I wrote this book based on my past 27 years' experience in the fields of strategic planning, execution, and measurement. This experience has been gathered partly from our client corporations and partly from our family business. Hence, this book presents both a consultant's view of strategic planning and also an entrepreneurial approach. The book takes you through the entire process of strategic planning, execution, and measurement.

The first chapter defines world-class status, vision, mission, business model and value chain. It states the importance of all these components in the process of strategic planning. The two most important prerequisites of the exercise of strategic planning are discussed in the second chapter. For successful strategic planning, understanding the present status of an organization is crucial. This requires a thorough inquiry into the present operations, systems, and structure of the organization. This chapter offers a comprehensive questionnaire for conducting such a detailed inquiry. It also gives a matrix of 25 parameters for assessing the status of the organization. This exercise helps to bring out the perceptional and aspirational gaps among the top executives and owners. An effective strategic plan requires these gaps to be bridged.

The structure of a strategic plan and the preparation process are discussed in detail in the third chapter. One feature of this chapter is the discussion of 150 generic strategies in five performance or growth areas. The fourth chapter describes the monitoring mechanism required for strategy execution. The fifth chapter presents the financial impact of strategy execution and explains various financial monitoring parameters used in the performance cards of the individual employees.

Chapter 6 explains an entrepreneurial approach to strategic planning. A strategic plan is basically a broad road map enabling an organization to reach the destination of entrepreneurial excellence. Hence, this chapter contains Entrepreneurial Score Cards (ESCs) for each type and category of employee across the organization.

The seventh and final chapter is a comprehensive case study on the preparation of a strategic plan and its conversion into an annual budget. The correlation between long-term strategies and relevant annual targets is demonstrated through this case study.

I cannot complete my preface without expressing my gratitude to certain people. I must thank my publisher for showing faith in me, and my clients who encouraged me to write this book. I thank my friend, Mandar Vartak, who gave me valuable suggestions. I acknowledge the efforts of my wife, Manjiri, who carried out the painful process of editing this book. I hope my readers will appreciate my efforts and theirs.

Girish P. Jakhotiya

Author

Girish P. Jakhotiya, MCom, FCMA, PhD,
is a renowned management consultant, educator, researcher, author, and speaker. He has conducted consultations with more than 60 multinational and Indian corporations and banks, including Siemens, British Petroleum, World Bank, ILO, Philips, Tata Group, Mahindra & Mahindra Group, Larsen & Toubro Group, Forbes Group, Hindustan Unilever, Castrol, Indian Oil, SICOM, IDBI Bank, Central Bank of Oman, Sesa Goa Group, and Dempo Group. He has also served as an advisor for such government organizations as NABARD and the Agriculture Finance Corporation, among others. He developed the Economic Growth Model for the government of Oman. Based on his consulting experience and research, Dr. Jakhotiya designed a comprehensive business management model titled AGNI (Aggregate Growth through Networking and Innovation). AGNI is patent pending.

Before venturing into a profession of management consulting, research and training, Dr. Jakhotiya was a senior professor of finance and strategy at the well-known Jamnalal Bajaj Institute of Management Studies, University of Mumbai (Bombay), where he was named the "All India Best Young Professor of Management" by the Association of Indian Management Schools (AIMS) in 1996. The Bombay Management Association designated him the "Best Management Teacher" in 1997.

Over the past 29 years, Dr. Jakhotiya has educated more than 27,000 corporate executives, government officials, entrepreneurs, and bankers in the areas of strategic management, financial management, public finance, and human resource management. He has also lectured on economic governance in special workshops organized for Indian politicians.

Under his able guidance, five scholars earned their doctorates. His training programs are based on contemporary ideas, live case studies, and relevant literature.

Dr. Jakhotiya's research focuses on fundamental issues and problems faced by the global economy. His original framework for this research is titled "Spiritual Economics" (or "Eternal Economics"). He maintains that a pragmatic and sustainable combination of socialist economic equity and capitalist entrepreneurial wisdom is the right answer for global dilemmas. Dr. Jakhotiya's management consulting firm Jakhotiya & Associates is known for offering innovative yet pragmatic solutions. Dr. Jakhotiya and his team strongly believe in acquiring intellectual property rights and using them appropriately for industry and society at large.

Dr. Jakhotiya is the author of five books: *Strategic Financial Management, Revised and Enlarged Edition* (Vikas Publishing House, Delhi), *Finance Made Simple*, coauthored with Manjiri G. Jakhotiya (Jakhotiya & Associates, Mumbai), *Portfolio Management: Scope for Management Accountants* (ACAI, Delhi), *Glossary in Management Accounting* (ACAI), and *Strategic Management Accounting* (ICAI).

Dr. Jakhotiya can be contacted by e-mail at girishjakhotiya@gmail.com or readers may visit his website (www.jakhotiya.com).

Chapter 1

Strategic Planning: A Tool for Reaching World-Class Status

Defining World-Class Status

A business organization achieves world-class status when it fulfills the following six eligibility criteria. A world-class organization:

- Is ready to compete globally in any market.
- Constantly strives to reach higher benchmarks.
- Regularly innovates its products, processes, or employee competencies.
- Enjoys competitive advantages in at least two of six value-driving areas (or segments): product, price, employee competence, process excellence, perception, and programs of change.
- Is one of the top three or four players that collectively dominate 40% of their markets.
- Retains its world-class status at least for a decade.

The process of attaining world-class status may start with a single product or process that meets the criteria. Such a modest beginning may trigger the long and complex process of total transformation.

Gaining World-Class Status

A business enterprise gains world-class status mainly through networking and innovation. Obviously, the organization must be entrepreneurial and work simultaneously on five essential growth and performance areas:

1. Business growth and expansion
2. Strategic cost management
3. Operations, logistics, and technology management
4. Employee development and satisfaction
5. Resource management

These five performing areas lead a business organization to balanced growth. An enterprise growing only through volume and failing to work on employee competencies required for the future may not be able to sustain or manage its volume growth over the long run. For this reason, an enterprise needs to focus systematically on the various components of these five growth areas:

Business Development
   Routine business development
   New business development
   Consolidated business development
Operational Improvement
   Quantitative operations management
   Strategic aspects of operations management
   Consolidated operations management
Cost Competitiveness
   Operations cost management
   Cost reduction and cost competitiveness
   Strategic cost management
   Application of activity-based costing
Productivity Enhancement
   Knowledge acquisition
   Knowledge application
   Strategic knowledge management
   Routine productivity follow-up
   Productivity enhancement efforts and results
Resource Facilitation
   Asset utilization efficiency

Routine capital expenditures
Special capital expenditures
Strategic fund mobilization and deployment

An enterprise must thrive continuously to sustain world-class status. It must therefore monitor and work on the six eligibility criteria defined above and outlined below:

Global Preparedness
Product
Cost
Delivery

Achievement of Higher Benchmarks
Institutionalization of benchmarking
Courage to challenge established benchmarks
Mastering language of benchmarking

Regular Innovation
Products
Processes
Employee competencies

Competitive Advantage
Dominance for at least two value drivers
Networking to gain competitive advantage
Strategic business modeling

Global Leadership
Part of market leadership
Sustainability
Brand ability

Decade-Long Maintenance of World-Class Status
Universal vision
Workable ideas
Strategic leadership

Figure 1.1 qualitatively illustrates the "horoscope" of a world-class corporation and the components are described in detail below.

Selectivity (focus)—Attaining world-class status requires constant work on an organization's sustainable core competence to eventually reach a focused area of business concentration. In other words, an organization may not become world-class if it ventures into too many areas and divides its

4 ■ *Strategic Planning, Execution, and Measurement*

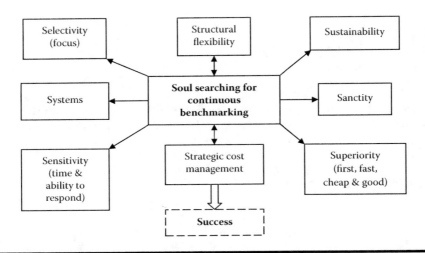

Figure 1.1 "Horoscope" of world-class corporation.

energy. Focus is also required to sharpen the attributes of an enterprise that ultimately produce a competitive advantage.

Structural flexibility—A world-class corporation must respond proactively to all its stakeholders on one side and to market challenges on the other. Therefore, the organization requires a flexible structure that can facilitate speedy decision-making processes and feedback mechanisms. In other words, a world-class company is also entrepreneurial across its various components. A complex structure discourages entrepreneurship because it creates bureaucracy.

Sustainability—Achieving world-class status may be easier than sustaining it. This is because many world-class corporations become lethargic or self-centered over time and therefore neglect the efforts required to retain their competitive advantages. To achieve sustainability, an organization must work rigorously to add new competencies and maintain a competitive gap between itself and its competitors. Sustainability also refers to the capacity of an organization to perform under any conditions. Thus, sustainability requires an innovation culture.

Effective systems—As noted earlier, leaders of a world-class corporation should focus on growth and development. Hence, they have little time to attend to routine business. Strong systems should be in place to facilitate flawless management of daily operations. Standard systems also facilitate management by objectives, better performance monitoring and measurement. The four major systems are manufacturing, marketing, human resources, and finance and control.

Sanctity—This component is defined as a transparent, reasonable, and adequate equilibrium between corporate governance and enterprise governance. Corporate governance consists of ethical business practices, whereas enterprise governance concerns entrepreneurial freedom. It is always a challenging task for a world-class organization to balance these two factors. Sanctity of purpose and processes is a very important aspect of retaining world-class status. Sanctity benefits everyone—shareholders, customers, employees, and other stakeholders. It reduces transactional cost and adds to the brand equity of an organization.

Sensitivity—We define sensitivity as alertness to the impacts of known and unknown business variables. It may also be defined as a combination of intellectual, operational, emotional, and entrepreneurial quotients of a business. Many great organizations lose their customer focus because they do not remain sensitive. Sensitivity inside an organization is also crucial to maintain productivity at all the levels of hierarchy.

Strategic cost management—Costs must be managed strategically, that is, they should be incurred appropriately based on proportionate value to be offered to customers. Strategic cost management is also a collective result of constant experimentation through innovative ideas, process improvisation, and employee performance. Cost management represents a vital competitive advantage and no world-class corporation can neglect the need to reduce costs sustainably. Accurate application of a combination of activity-based costing and objective-based costing is a prerequisite for utilizing costs to gain a competitive advantage.

Superiority—We assume that a world-class corporation achieves superiority on all four product characteristics: first, fast, cheap, and good. In other words, a world-class operation must maintain its superior status by entering a new market or working on a new idea first. It must execute commercially viable ideas quickly to create a competitive gap. A product must be cheap to respond effectively to customer economics. Naturally, all these three advantages are admired by customers but a company must also offer the final feature: a good product or service.

Improving benchmarking—Continuous benchmarking—a genuine attempt to achieve higher and better parameters of performance—must become the culture of a world-class organization. Entrepreneurial results demand constant work on better benchmarks because this improves the ability to maintain a competitive market advantage. Better benchmarks may be obtained from various sources such as market data, analysis of past performance trends, shareholders, and other industries. Continuous benchmarking should be facilitated through a performance management system. Top executives of an organization should set examples by working on higher benchmarks so that they may expect (and receive) similar responses from their juniors.

Benefits of Strategic Planning

Reaching world-class status requires both a clear vision and a systematic approach to pursuing the vision. This systematic approach or (broad road map) is known as a strategic plan that consists of several components discussed in detail in Chapter 3. First we should understand the advantages of strategic planning. It:

1. Clarifies the certainties (and uncertainties) on the path to reaching the vision and details the milestones of the journey to the desired destination.
2. Should be based on detailed scenario mapping and describe the various types and levels of sensitivities found in business environments.
3. Describes various hurdles expected during strategy execution; a proactive approach to managing obstacles is desirable.
4. Acts as a comprehensive set of performance-based strategies by spelling out the complementary and contradictory relationships among various functional strategies. This understanding of relationships helps predict the ultimate result of combining strategies.
5. Serves as a long-term (3 to 5 years) road map and thus becomes a vital input for resource planning. Simultaneously and/or interdependently executed strategies require different resources at different times. An effective strategic plan clearly prescribes such resource requirements.
6. Presents specific execution and impact routes for all strategies. An execution route should describe in chronological order the tasks required to implement a strategy successfully. An impact route should define the long-term commercial and/or strategic impacts of each strategy on the business. Both routes should facilitate performance budgeting and financial budgeting. In other words, a strategic plan serves as a foundation for preparing the annual budgets that are essential for evaluating performance.
7. Details exit routes for use in the event a strategy fails. A good strategy may prove ineffective during execution. For that reason, every strategy should have a damage control mechanism. This is particularly true for strategies that may be subject to a number of complex uncertainties during their execution.

8. Covers 3 to 5 years so that a failure in the first year can be overcome in the second year through strategy revival. In other words, a strategic plan motivates a business team by spreading performance assurances and scopes for revival over a span of 3 to 5 years. If a business enterprise does not achieve its desired return on investment (ROI) in the first year, it may revise the plan and recover in the second and third years and enable the enterprise to reach its desired average ROI despite the setback.

Fundamental Factors Defining Scope of Strategic Planning

As stated earlier, every CEO cherishes the dream of attaining world-class status for his or her organization. The dream can be realized by following a systematic road map called a strategic plan. In other words, the CEO's dream is a fundamental reference that decides the overall scope of strategic planning and is called the organization's vision—the basic input that leads to entrepreneurship.

Vision may be defined qualitatively or quantitatively or both ways. An organization's vision should be understood by all its internal and external stakeholders and achieved through the definitions of functional or segmental visions. Of course, the convergence of all these segmental visions should constitute the total vision of the organization. A vision is necessary for providing:

Reason to perform—A vision clearly defines a holistic purpose for all the stakeholders to perform in one direction and with a single clear ambition. Without a well-defined purpose for reaching a clear destination, the journey of becoming a world-class entity could proceed without control or measurement or fail completely. Vision inspires all employees and functions across an organization.

Visibility—A business organization is known by its vision. People admire and accept the quality of leadership if the statement of vision is well prescribed. Business partners will show a willingness to join if they appreciate the visibility of an organization. A vision offers such visibility.

Energy to succeed and sustain—When a destination is known and assured, a very long and tiring journey also becomes enjoyable. This applies to every ambitious and successful business organization. Its stakeholders create and preserve their energy to ensure they are adequate for the difficult journey. Ups and downs create pressure on available resources. A clear-cut vision keeps the terminal point tangible so that cyclical fluctuations do not hamper the confidence of the stakeholders.

Framework for mission—A clear vision defines a set of major performance measures. In other words, a comprehensive mission can be decided in the light of a vision and the mission becomes a set of goals and objectives for the organization to pursue. Every stakeholder becomes a member of a strong team that is ready to take on the mission.

Corporate homogeneity—All business verticals and functional components can achieve a homogeneous ambition and approach if their aspirations and perceptions are well guided. A vision document provides the required homogeneity. All stakeholders can perform with better zeal if they connect to a common purpose provided by the organization's vision.

In addition to defining the vision of an enterprise, its CEO should understand the factors that may impact or restrict the vision:

Lack of ambition
Lack of exposure
Lack of courage
Limited opportunities
Limited resources
Limited competence
Defining future as uncertainty
Individualistic performances

Institutionalization of Vision and Mission

The corporate vision should be institutionalized. That means that every employee (at every level) of the organization should understand and own it in an effort to create an overall homogeneity in understanding its purpose and future. The vision of the organization should be translated further into a vision for each functional department and business vertical. For example, the human resources (HR) function should have its own unique vision statement. If the HR function presently has a facilitative role, its vision should be to reach a strategic or business partnering role in 3 to 5 years.

Let us see an example to understand business vertical vision. A company may consist of three business verticals: infrastructural construction, energy, and industrial goods. It should have a separate vision statement for each vertical. For example, the vision statement for the energy vertical may be "reaching world-class status with affordable prices of energy to customers

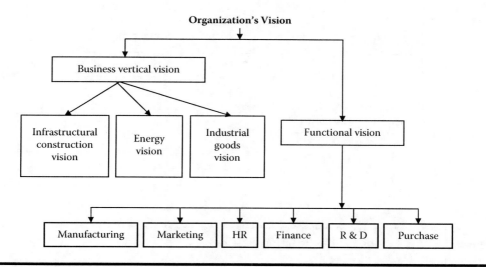

Figure 1.2 Organization's Vision.

and a commitment to work on renewable energy." As stated earlier, the organization's vision should decide or guide the visions for functions and verticals. See Figure 1.2.

It may be possible to define a vision applicable to each factory within a manufacturing function by delineating certain rates of maturity or performance to be attained over a set period of time.

Functional vision and business vision are useful concepts for describing the details of strategic plans to be applied to each function and business. Of course, all functions and businesses are ultimately the organs of an organization therefore there should always be parts of an integrated strategic plan covering the entire organization.

Vision statements for support functions like HR, accounting, and research and development (R&D) are as important as those for operating divisions. The employees working in these areas should clearly understand their roles and purposes and feel that they too fill important functions within the organization.

A vision is achieved through an appropriate mission. A mission consists of a set of the most important performances or drives of an organization. As noted earlier, a mission should contain statements of commitments in five performance areas (business growth, cost management, employee development, operational and technology management, and resource management). The CEO and senior management team should raise the questions cited in Figure 1.3 in determining the organization's mission. See Figure 1.3.

The mission components should be specific for the vision and the mission should operate like a vehicle that allows the organization to reach

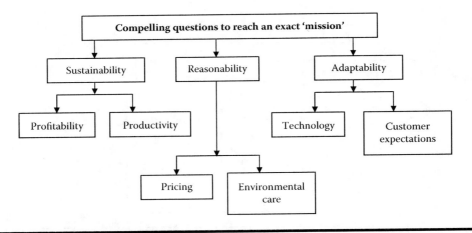

Figure 1.3 Questions for determining mission.

the destination defined by its vision. Therefore, the mission should be tested by answering certain questions:

1. Will the mission be sustainable in terms of profitability and productivity expected from the business model over the long run?
2. Will the mission reasonably offer appropriate value at appropriate cost and also take care of the external environment?
3. Will the mission be adaptable, that is, will it be flexible enough to absorb and apply changing technology and fast-changing customer expectations?

Answers to these questions should allow an organization to assess the overall effectiveness of the mission statement quickly and modify the statement as required. To illustrate these concepts, we will consider a simple example of correlation of a vision and mission.

Example 1.1: Indian Automobile Company

The company's vision is "to be a world-class automobile company by 2016, with the highest market share in Asia." The mission steps necessary to achieve the vision are to:

- Improve product quality and cost competitiveness based on global benchmarks.
- Develop global partnerships to enlarge the network in Europe, North America, and Asia.

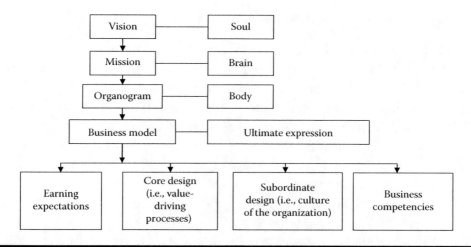

Figure 1.4 Logical flow for achieving vision.

- Strengthen internal performance systems to institutionalize entrepreneurship at all employee levels.

To summarize, the vision is achieved through the mission. Figure 1.4 illustrates the logical flow for attaining an organization's vision.

Other Important Factors: Business Model, Core Competence, and Value Chain

Execution of a broad mission requires a clear business model (BM) that combines hard and soft aspects of a business organization. In other words, a business model represents a balanced convergence of earning expectations, value-driving processes or elements, organizational culture, and various competencies. The choice of a business model is mainly decided by several factors including:

- Organization's vision and mission
- External factors such as government policies, market forces, and types of customers
- Organizational ethos
- Core competence
- Technical and operational factors that determine scope of the value chain
- Entrepreneurial styles and strengths of top executives
- Related factors such as socioeconomic conditions, environmental issues, etc.

12 ■ *Strategic Planning, Execution, and Measurement*

Business Model and Return on Investment

Figure 1.5 illustrates five basic components of a simple business model. In many organizations, earning expectations decide the business model. Other factors like ethos, competencies, and processes become secondary.

Benchmarking earning expectations (also known as return on investment or ROI) is a vital activity of strategic planning. A CEO with the support of the CFO and COO should benchmark rates of returns on investments carefully because ROI normal takes center stage on a strategic plan. Figure 1.6 depicts the five dimensions of ROI.

A strategic plan spanning 3 to 5 years should accommodate all the adjustments required to meet the rate of ROI to be earned during both good and bad times. In other words, a CEO should aim pragmatically for an average ROI over

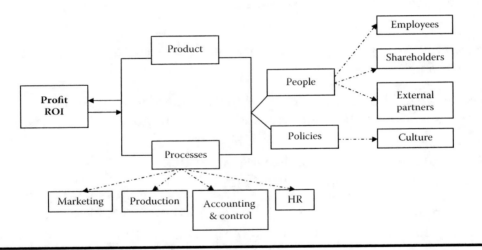

Figure 1.5 Five components of business model.

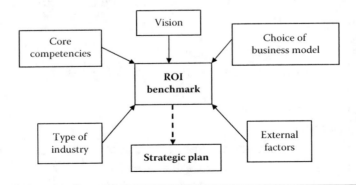

Figure 1.6 Dimensions of Return on Investment (ROI).

3 or 5 years, as the case may be. Figure 1.7 illustrates a strategic plan for achieving average ROI. Figure 1.8 shows the four levels of benchmarking ROI.

Benchmarking of the ROI at the economy level is very crucial because it serves as the base ROI for defining further benchmarks. The CEO and CFO must realize that their national economy will influence their ROI aspirations. An error in benchmarking may create serious flaws in the ensuing strategic plan. The bottom-up approach of ROI-based strategic planning is shown in Figure 1.9.

Let us look at the scientific approach of benchmarking an economy level ROI. Table 1.1 compares economy level ROI factors in India and England. The structure of an economy level ROI is dynamic because interest and inflation rates vary by location and over time. For that reason, an average

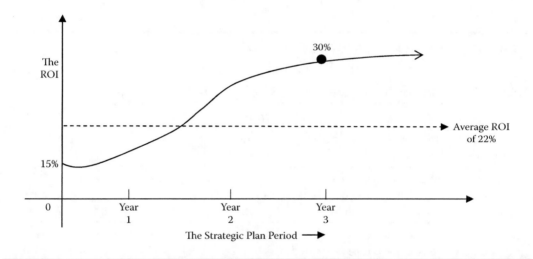

Figure 1.7 Strategic plan for average ROI.

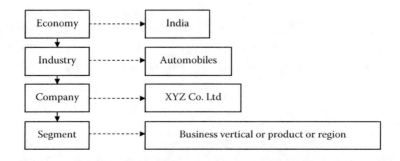

Figure 1.8 Four levels of benchmarking.

14 ■ *Strategic Planning, Execution, and Measurement*

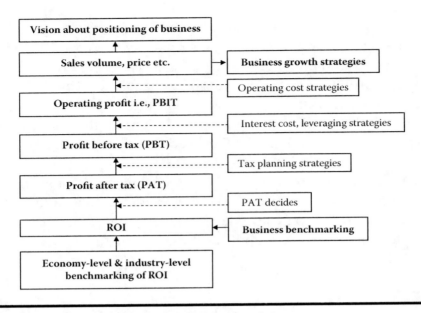

Figure 1.9 Bottom-up approach of strategic planning.

Table 1.1 Comparison of Economy Level ROIs in India and England

ROI Component	India (%)	England (%)	Explanation
(a) Interest rate or rate of inflation (whichever is higher)	12.0	5.0	→ Cost of capital
Add (b) real, disposable, or consumable income [should be 40% of component (a) considering local cost of living]	4.8	2.0	→ Profit withdrawal
Add (c) growth rate (may be moderately set as lowest rate of growth in economic sector, e.g., agriculture)	3.0	1.0	→ Reinvestment of profit into business
Total	19.8	8.0	

ROI spanning 3 to 5 years should be considered in compiling a strategic plan. The concept of averaging ROI allows a company to accommodate the normal ups and downs (booms and recessions) of an economy over time.

Core Competence

Another vital aspect of preparing a strategic plan is core competence—a concept based on relativity. Core competence is simply the ability to acquire various skills and use them successfully for different business purposes in different regions. Normally, it is presumed that an organization (or entrepreneur) should select a business or decide a business model based on available core competence. Defining a vision based on core competence takes the concept one step further.

A company should restrict its vision within the scope allowed by its core competence. In a globalized scenario, core competence can be acquired or hired. A group of companies that pursue different businesses require and utilize core competencies relevant to their operations. A diversified group must develop and use core competencies based on the requirements of conducting diversified businesses. Diversification reduces the overall risk of the business portfolio of a group.

Therefore, a strategic plan may be based on an existing core competence or it may plan to acquire a new competence. Worldwide businesses achieve growth through strategies of acquiring diversified competencies. In other words, a part of the strategic plan should be devoted to issues related to acquiring new competencies. Core competence is simply an expertise of a company that produces best results.

Core competence must offer a sustained competitive advantage. Over a period of time, competitor successes may threaten the competitive advantage of a business enterprise. Therefore, an enterprise must evolve new strategies to improve its existing competitive advantage or gain a new advantage. A business enterprise must constantly work to upgrade its core competence by analyzing performances in the five growth areas mentioned earlier by following its strategic plan.

Value Chain

Let us now discuss the value chain—a consecutive expression of all value-adding processes (or activities) conducted via a clear business model based on a benchmarked ROI projection. A value chain may consist of outsourcing and insourcing activities based on the organization's core competence and the competitive advantages produced by insourcing and outsourcing. See Figure 1.10, which shows value-adding processes (1 through 5). A value chain should be analyzed to ensure operational convenience, financial viability, and strategic (or long-term) sustenance.

16 ■ *Strategic Planning, Execution, and Measurement*

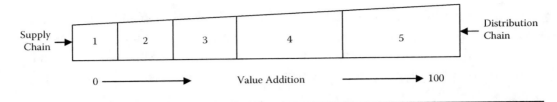

Figure 1.10 Value chain.

Value, also known as utility, is a critical factor in every business operation. A product (or service) may be expected to offer one or more of the following values to customers:

Real value: genuine utility offered by a product through its attributes.
Esteem value: a "premium" over and above real value offered to customers at an extra cost to satisfy customers' desire for esteem.
Perceived value: utility perceived by customers may be different from real utility; the result is a perceptional gap created by a communication gap; perceived value may be dangerous.
Practical value: the utility a customer expects and can buy.

A business enterprise must carefully choose the values it offers its customers. It may offer different values through different products to different customers in different markets. Value represents the combination of vision for growth, retention of competitive advantages, adherence to business models, and other factors.

A business enterprise may occasionally have to conduct a strategic exercise known as value appropriation—adjusting the value offered to customers so that it can retain its market share or leadership. Value appropriation is different from value dilution and value engineering. It is a strategic attempt to offer practical value to customers as shown in the following illustration.

Example 1.2: A Company and B Company

These two companies are competitors. Both sell low-quality fans in a down market. Figure 1.11 shows a comparison of both companies' products. B Company started losing its market share to A Company because of a price differential of $2 per fan arising from price elasticity due to demand in

Strategic Planning: A Tool for Reaching World-Class Status ■ 17

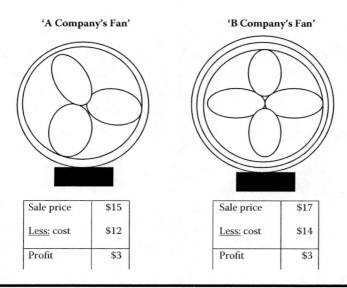

Figure 1.11 Comparison of fans manufactured by firms cited in Example 1.2.

Table 1.2 B Company's Value Appropriation Exercise

Cost Saving (Value Appreciating) Factor	Cost Saved ($)
Reduction from four blades to three based on air circulation expected by customers	0.60
Reduction from three safety rings to two based on safety expected by customers	0.90
Changes in paint and painting process	0.30
Change of fan base from steel to durable plastic	0.70
Overall cost saving	**2.50**

a down market. B Company decided to carry out a strategic value appropriation exercise to adjust its price. The details are shown in Table 1.2.

If B Company decides to transfer this cost advantage to its customers, the sale price would be $14.50. This price should be competitive enough to allow B Company to regain and even increase its market share. The company's value appropriation exercise was intended to offer practical value to its customers at an affordable price.

Let us now consider a balanced and strategic combination of vision, business model, value chain, and organizational structure.

Example 1.3: ABC Company

This company is a market leader for skin-care products and operates 40 factories in industrially backward areas in different parts of the country. The factories are either franchise operations or fully owned subsidiaries. ABC owns a few strong brands and spends heavily to retain its earning power. Its factories are state-of-the-art production centers. The production processes are simple and production costs are far less than marketing and branding costs (see Figure 1.12).

Obviously ABC's core competence is marketing, which provides sustained competitive advantages based on brand equity. Because ABC operates factories in economically backward areas, it gains the advantages of tax holidays, subsidized power and water, and other benefits. In return it must employ 50% of factory workers from tribal communities. This is effective because ABC's production processes are simple.

Each ABC factory is run by five managers (general manager, production manager, accounts manager, personnel manager, and quality manager). They are the only factory employees on ABC's direct payroll. These managers work at an assigned factory for 3 years and then move to another factory. The permanent factory employees are production and office workers who are on the payrolls of the relevant franchisees or subsidiaries. ABC pays local- or region-based salaries to these permanent local employees; their salaries are much lower than those of ABC's direct (management) employees. Thus, ABC saves a lot of payroll costs. The savings are essential because ABC must spend much more to maintain its marketing function.

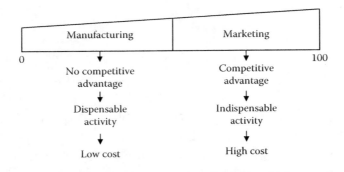

Figure 1.12 Comparison of performance of manufacturing and marketing functions based on discussion in Example 1.3.

Strategically, ABC balances its competitive advantage and cost-saving operations. A franchise is a simple model of outsourcing a portion of the value chain but it is insignificant in terms of sustainable competitive advantage.

ABC's vision is simple. It wants to be an unchallenged market leader for most of its products that offer a variety of value propositions to customers. ABC's business model is based on the principles of high volume and low margin. Therefore, it must work constantly to grow its market share by analyzing the performances of its manufacturing and marketing functions (Figure 1.12). ABC assesses the performance of its factories using only three parameters: volume, quality, and cost.

Chapter 2

Prerequisites of Strategic Planning: Inquiry and Status Evaluation

Prerequisite 1: Examining Performances, Perspectives, and Aspirations

Strategic planning should be based on the thorough knowledge of the present and potential strengths and weaknesses of an organization. Planners should consider the possible opportunities and future threats. In other words, present performance of various business verticals and functional departments should be thoroughly known by those who are responsible for conducting strategic planning exercises. Obviously, the top executives are expected to take the lead in framing strategies for the five performance areas mentioned in Chapter 1.

However, these executives often lack the time and appropriate methodologies to strategically examine their own performances, perspectives, and aspirations about their organization and may lack the confidence to pursue their visions along the broad road map called a strategic plan. The best approach is for top management to carry out a thorough strategic inquiry into five major areas of business performance:

- Business development
- Operations management
- Strategic cost management

Employee productivity
Strategic resource management

Each top executive should answer a versatile questionnaire covering various performance areas. The answers constitute a self-inquiry that facilitates a systematic analysis of executives' perspectives and aspirations. This method also offers a systematic approach to analyzing an organization's strengths and weaknesses if the executives share their responses to the questionnaire and reconcile their perceptions.

Strategic inquiry serves as an input for deciding the organization's status at the start of the strategic planning exercise. A complete strategic inquiry reveals valuable data by allowing management a microscopic view of the organization's performance. It stimulates every top executive to think alternatively, innovatively, and comprehensively about the organization's strengths, weaknesses, opportunities, and threats.

The next section is a series of comprehensive questionnaires for facilitating a strategic inquiry. The inquiry has been divided into six parts and is flexible enough to suit any organization of any size in any industry. Senior executives and their subordinates should try to answer the questionnaires carefully and completely by a defined deadline—usually a month. Each question refers to the table that follows and the purpose of the question is indicated.

Part A: General Inquiry

1. Which items listed in Table 2.1 are the competitive and strategic advantages of your organization? [Purpose: Assessment of competitive and sustainable strength of the organization in comparison with competitors.]

Table 2.1 Competitive and Sustainable Strength

No.	External Advantages	No. 1 in Industry	No. 2 in Industry	In Top Three	Close to Average	Negligible
1	Product quality					
2	Distribution network					
3	Supply reliability (time and volume)					
4	Price reasonability					
5	Brand image					
6	Cost of funding					
7	Empowerment					
8	Strategic locations					
9	First entry advantage					
10	Technology					
11	Employee support					
12	Economy of scale					
13	Collaborative advantage					
14	Fiscal advantage					
15	Written-off cost advantage					
16	Fund availability					
17	Big brothers' shelter					
18	Macrolevel understanding and vision mapping					
19	Historically acquired assets					
20	Cumulative research base					

2. What advantages listed in Table 2.2 does your organization present to the parent group? [Purpose: Evaluation of strategic, operational, and financial advantages of the organization to company group.]
3. What should your company's future strategic positioning be based on regarding the parameters listed in Table 2.3? [Purpose: Visualization of strategic position of the company based on structure, business partners, resource requirements, and growth and development factors.]

Table 2.2 Evaluation of Strategic, Operational, and Financial Advantage

No.	Group Advantages	Very Considerable	Considerable	Negligible	Nil
1	Product focus and strategic business units (SBU) advantages				
2	Enriching group's value chain				
3	Contribution to group's profits				
4	Contribution to group's strategic plans				
5	Role-model positioning				
6	Center for initiatives				
7	Administrative diversification				
8	Synergistic advantage				
9	Financial engineering for net worth appreciation				
10	Group's structural flexibility				

Table 2.3 Visualization of Strategic Position of the Company

No.	Parameter	Yes/No
1	Your company should be totally independent of its holding company, i.e., the holding company should be an investor only; ownership and management should be separate	
2	New partners in your company should contribute: Capital Borrowed funding Equity funding Technology upgrades New markets Other advantages	
3	Your company should conduct certain activities as a member of a consortium, cartel, confederation, or flexible group: Procurement of raw material Product processing Product distribution Product development Fund-raising New market development Wage negotiations Human resource development	
4	Your company should pursue certain development- and growth-oriented changes: Introduction of high quality products Capacity expansion Equity sharing with employees Creation of distribution network Building brands Acquiring strong brands Creation of operational franchises Offering promotional and/or consultative support Product diversification	

4. The financial department should complete Table 2.4 for strategic analysis of product-wise financial performance. [Purpose: Financial analysis of product performance in comparison to best and average products in the industry.]

Table 2.4 Financial Analysis of the Product-wise Performance

Detail	Product 1			Product 2		
	Your Company	Best in Industry	Average in Industry	Your Company	Best in Industry	Average in Industry
Identifiable cost per unit						
Add common cost allotted per unit						
Total cost per unit						
Add profit per unit						
Sale price per unit						
Return on investment (ROI)						
Investment turnover (sales/investment)						
Profitability (profit/sales)						

5. To rate organizational effectiveness, rank the value drivers (factors contributing to organizational effectiveness) in Table 2.5 in their order of importance based on their actual contributions to organizational success. [Purpose: Assessment of the impacts of various value drivers on the effectiveness of the organization.]

Table 2.5   Impact of Various Value-Drivers

No.	Value Drivers	Rank in Order of Importance	Rank by Actual Contribution to Organization Success
1	Processes (technology and operational smoothness)		
2	Products (quality, volume, reliability)		
3	Price to customers		
4	People (quality and availability of employees)		
5	Promotion (promoting new products and capturing new markets)		
6	Policy (informal rules, code of conduct, etc.)		
7	Places (locations of production, distribution, marketing, and other activities)		
8	Programs (routine and special activities or projects for promoting development and growth)		
9	Power of funds (amounts and costs)		
10	Pragmatism (flexibility and rationality)		

6. To rate performance management, rank the factors listed in Table 2.6 in terms of their importance for performance planning. [Purpose: Understanding the importance of various factors considered in performance planning.]

Table 2.6 Importance of Various Factors in Performance Planning

No.	Factor	Rank
1	Long-term goal of organization	
2	Present resources available	
3	Competence of employees	
4	Projected threats from competitors	
5	Government rules and regulations	
6	Cost advantage	
7	Quality enhancement	
8	Gaining greater market share	
9	Rate of return expected on owners' funds	
10	Quality of internal logistics	
11	Organizational culture	
12	Techniques available for performance monitoring	
13	Immediate priorities for change or improvement	

7. As a division, department, or function head, which parameters or guidelines listed in Table 2.7 do you use for performance planning? Answer yes or no, and rank each item based on its importance. [Purpose: Understanding various parameters of performance planning at the division or department level.]

Table 2.7 Parameters of Performance Planning at Division or Department Level

No.	Parameter	Yes/No	Rank if Yes
1	Targets set by top management or corporate planning cell		
2	Specific targets allotted to you		
3	Actual performance of previous year		
4	Performance of similar department in competitor organization		
5	Ideas for future growth of your department		
6	Competence and size of your team		
7	Present resources available		
8	Trend analysis covering past few years		
9	Notional earning or cost benchmarks		
10	Feedback from subordinates		
11	Strategic importance of your department within organization		
12	Other parameter (specify)		

8. Which of the methods, processes, or parameters shown in Table 2.8 do you use for performance monitoring? Answer yes or no, and rank each yes item based on degree of importance, convenience, and transparency. [Purpose: Understanding various sources of performance benchmarks and their use in performance control.]

Table 2.8 Sources of Performance Benchmarks

No.	Performance Monitoring Technique or Parameter	Yes/No	Rank Yes Answers by		
			Importance	Convenience	Transparency
1	Quantitative targets				
2	Competitors' benchmarks				
3	Customer feedback				
4	Supplier feedback				
5	Employee feedback				
6	Qualitative targets				
7	Surprise checks				
8	Trend comparisons				
9	Global standards				
10	Informal information channels				
11	Self-assessment reports				
12	Early warning signals				
13	Notional parameters				
14	Industry average performance				
15	Benchmarks set by parent organization				
16	Benchmarks set by shareholders				
17	Other technique or parameter (specify)				

9. Rank (by degree of significance) the components or features listed in Table 2.9 that you consider when framing an optimal strategy. [Purpose: Understanding various factors defining an optimal strategy and assessing their relative importance.]

Table 2.9   Defining an Optimal Strategy

No.	Component or Feature	Rank
1	Sustainable results	
2	Simplicity of interpretation	
3	Ease of implementation	
4	Collective participation in formulation	
5	Collective participation in implementation	
6	Ease of assessing impact	
7	Value of cost or degree of risk	
8	Support from senior management and/or owners	
9	Scope for new competitive advantage	
10	Retention of present competitive advantage	
11	No complex data requirement for formulation	
12	Other parameter or technique (specify)	

10. In your organization, what actual preference is given to strategy quotients in Table 2.10? Indicate your optimal preferences in the right-most column. [Purpose: Appreciating the application of various quotients in strategic planning.]

Table 2.10 Quotients in Strategic Planning

No.	Strategy Quotient	Actual Preference in Organization	Optimal Preference (Your Opinion)
1	Entrepreneurial		
2	Intelligence		
3	Economic		
4	Emotional		
5	Social		
6	Operating		
7	Leadership		

Prerequisites of Strategic Planning: Inquiry and Status Evaluation ■ 33

11. Assess the quality and quantity of corporate governance in your organization based on the parameters listed in Table 2.11. Rate each parameter excellent, adequate, or inadequate. [Purpose: Understanding and evaluating parameters of corporate governance.]

Table 2.11 Parameters or Components of Corporate Governance

No.	Parameter	Quality and Quantity of Corporate Governance		
		Excellent	Adequate	Inadequate
1	Accounting transparency			
2	Participation by minority shareholders			
3	Optimum employee utilization			
4	Optimal response to social environment			
5	Strategic inputs by board members			
6	Operational control by board members			
7	Decision-making efficiency and effectiveness of board			
8	Board teamwork			
9	Board responses to external stakeholders (lenders, customers, suppliers, etc.)			
10	Optimal dividend payments			
11	Board support to senior employees			
12	Optimal managerial remuneration			
13	Regularity, transparency, and effectiveness of board meetings			
14	Democratic system for discussions by working groups on major issues and decisions			
15	Ability of system to appraise quality and quantity of corporate governance			

Part B: Business Development Inquiry

1. Rate the expected future performance of your products (excellent, good, poor, or nil) based on the variables listed in Table 2.12. [Purpose: Assessing potential performance of company products in the future.]
2. Which of the threats shown in Table 2.13 do you expect your company to face in the coming years? Indicate the degree based on the key shown at the bottom of the table. [Purpose: Evaluation of potential threats expected in the near future.]
3. How do you rank the components of your business model shown in Table 2.14 on the key parameters of importance and scope for change? [Purpose: Assessment of components of the company's business model.]
4. Does the importance of each component of the business model change at various stages of business development? Rank the components listed in Table 2.15 in terms of importance in each of the seven stages shown. [Purpose: Critical evaluation of behaviors of various components of the business model at various stages of a company life cycle.]
5. How do you rank the business models used for various divisions in your organization based on the factors listed in Table 2.16? [Purpose: Comparative assessment of business models using three major parameters.]
6. What are the overall impacts of different business models used for different businesses in your group as defined in Table 2.17? [Purpose: Assessment of positive and negative impacts of business models.]
7. Which is the most important or vibrant subcomponent among those listed in Table 2.18 that considerably influences the overall design of your business model? [Purpose: Microanalysis of subcomponents of business model.]
8. How would you rank the business development strategies listed in Table 2.19 at various stages of your organization's life cycle? [Purpose: Appreciating the importance of adopting various strategies during different stages of a company's life cycle.]

Table 2.12 Assessment of Future Performance of the Company's Products

No.	Parameter	Excellent	Good	Poor	Nil
1	Market growth in next 3 years: Domestic Global				
2	Potential for growth in market share: Present rate Higher rate				
3	Potential for new or alternate product development				
4	Cost escalation				
5	Entry of new participants into market: Domestic Foreign				
6	Scope for new process and distribution technology development (and subsequent labor reduction)				
7	Need for capital investment for: Expansion Renovation Diversification				
8	Change in market forces based on: Price war Liberal credit Liberal commissions Distribution efforts Need for brand equity Increased volume of operations Quality expectations				

Table 2.13 Evaluation of Business Threats

Threat	Degree*
New entrants to market	
Product replacement	
Technology obsolescence	
Fund scarcity	
Reduced product demand	
Reduced economic abilities of customers	
Increasing cost of funds	
Reduced employee productivity	
Reduced empowerment by parent company	
Impact of parent company's downward performance	
International adversities	
Price and credit wars conducted by competitors	
Overall industrial slowdown	
Growing confusion in parent company's plans and strategies	

*0 = Nil, 1 = Little, 2 = Marginal, 3 = Considerable, 4 = High, 5 = Very high.

Table 2.14 Assessment of Business Model

No.	Component	Importance	Scope for Change
1	Profit (rate of return expected)		
2	Product		
3	Processes		
4	People		
5	Policies (work ethos)		

Table 2.15 Behavioral Evaluation of Business Model

Component	Entry	Take-off	Consolidation	Leadership	Diversification	Boom	Recession
Profit							
Product							
Processes							
People							
Policies							

Table 2.16 Ranking Business Model of Different Divisions

No.	Business Division	Ranking Factors		
		Risk	Rate of Return	Absolute Amount of Return
1				
2				
3				

Table 2.17 Impact of Business Models

No.	Impact	Agree to				
		No	Some Extent	Reasonable Extent	High Extent	Very High Extent
1	Diversity of business models creates hedge for risk management.					
2	Diversity has damaged homogeneity of overall group structure.					
3	Holistic and common strategic thinking among business heads is poor.					

Table 2.18 Microanalysis of the Subcomponents of the Business Model

Component	Subcomponent				
Profit	Operating profit	Profit after interest, before tax	Profit after tax	Cash profit after tax	Economic value added
Process	Manufacturing	Marketing	Accounting	Communication	Sourcing
Product	High value	Average value	Low value	Esteem value	Practical value
People	Entrepreneurial employees	Systemic employees	Fund providers	Distributors	Supply chain members
Policy	Customer handling	Government management	Environment care	Owner–manager relationship	Minority shareholders

Table 2.19 Ranking of Business Development Strategies

No.	Business Development Strategy	Life-Cycle Stage				
		Beginning	Take-off	Consolidation	Leadership	Weakness or Stagnation
1	Growth in volume					
2	Growth in product variety					
3	Partial or total diversification					
4	Backward integration					
5	Forward integration					
6	Growth via networking or partnering					
7	Creation of new markets					
8	Revival of old brands and products					

Part C: Strategic Cost Management Inquiry

1. Table 2.20 lists factors for achieving cost competitiveness and improving profitability. Rate the factors based on the key below the table.
   [Purpose: Understanding and evaluating methods to help company remain competitive and improve profitability.]
2. What weight do you attach to cost competitiveness compared to the other elements of competitiveness cited in Table 2.21?
   [Purpose: Comparison of the importance of cost competitiveness to other elements of competitiveness at various stages of a product's life cycle.]
3. How do you determine the benchmarks of cost competitiveness among those listed in Table 2.22? [Purpose: Understanding the relevance of various benchmarks of cost competitiveness.]

Table 2.20 Achieving Cost Competitiveness

Scope Provider	Scope*
Harsh supply negotiations	
Wages based on performance of individuals and departments	
Wages based on profits	
Process reengineering	
Product reengineering	
New product development	
Reduction in head office overhead	
Leasing out idle capacities	
Outsourcing processing and/or packaging	
Further automation	
Change in business model	
Capital restructuring	
Capacity expansion	
Backward integration	
Forward integration	
Brand use royalties	
Collaborative procurement	
Collaborative processing	
Collaborative distribution	
Reduction in decision making and implementation delays	

*0 = Nil, 1 = Little, 2 = Marginal, 3 = Considerable, 4 = High, 5 = Very high.

4. Compare the performances of your company and its competitors based on cost and other elements of competitiveness cited in Table 2.23. [Purpose: Assessing competitiveness of your company and its competitors.]
5. Table 2.24 lists various methods of measuring cost competitiveness. How do you measure cost competitiveness for your company? [Purpose: Understanding several methods of measuring cost competitiveness at various time intervals.]

Table 2.21 Other Elements of Competitiveness

Competitiveness	Product Performance Stage				
	Launch	Take-off	Consolidation	Leadership	Revival
Quality					
Size					
Delivery					
Cost					
Variety					
Sustainability					
Ethical practice					

Table 2.22 Benchmarks of Cost Competitiveness

No.	Benchmark	Cost Competitiveness				
		Product	Process	Period	People	Place
1	Market reference					
2	Trend analysis					
3	Earning or price target					
4	Unrelated industry benchmarks					

Table 2.23 Comparison of Cost Competitiveness

Element of Competitiveness	Competitors				
	Close	Best in Local Market	Local Average	Global Best	Global Average
Cost					
Quality					
Size					
Delivery					
Variety					
Sustainability					
Ethical practice					

Table 2.24 Methods of Measuring Cost Competitiveness

No.	Method of Measurement	Periodicity				
		Daily	Short Term	Medium Term	Long Term	Occasional
1	Per unit of product, process, or resource					
2	Percent of total cost					
3	Percent of total sales					
4	Discounted value (increase in NPV)					
5	Notional value (considering opportunity cost)					
6	Change in ROI					
7	Risk-weighted absolute cost saving					

6. Five methods of analyzing cost competitiveness are listed below. How does your operation track and treat cost competitiveness? [Purpose: Understanding the accounting or strategic treatment of cost competitiveness.] The analysis methods are:
   a. Accounting via product or process costing
   b. Separate accounting system that may not transfer cost advantages to customers
   c. Utilizing cost advantage to absorb future cost escalations
   d. Utilizing cost advantage to support weak product or territory
   e. Utilizing cost advantage notionally for project comparison
7. Which group among the stakeholders listed in Table 2.25 makes the greatest contribution to your company's cost competitiveness? [Purpose: Critical assessment of cost competitiveness contributions by various stakeholders.]
8. Which of the accounting and information systems listed in Table 2.26 do you use to capture adequate, reliable, and timely data about cost competitiveness? [Purpose: Appreciating various systems for capturing reliable cost competitiveness data.]

Table 2.25 Contribution to Cost Competitiveness

No.	Category	Contribution to Cost Competitiveness				
		Occasional	Sustainable	Always High	Nil	Negative
1	Employees: Entrepreneurial Knowledge Systemic					
2	Vendors					
3	Strategic partners					
4	Distributors					
5	Lenders					
6	Shareholders					
7	Government entities					
8	Advisors					
9	Others					

Table 2.26 Accounting and Information Systems to Capture Cost Competitiveness Data

No.	Accounting and Information Systems	Use			
		Regular Official	Occasional	Regular Unofficial	Supplementary
1	Profit center				
2	Activity-based costing				
3	Objective-based costing				
4	Segmental accounting				
5	Notional costing				
6	Incremental cost–benefit analysis				

Part D: Operations Management Inquiry

1. How do you evaluate your company's organizational structure in terms of the indicators listed in Table 2.27 on a decreasing scale of 5 (excellent) to 0 (nil). [Purpose: Critical assessment of the effectiveness of management.]

Table 2.27 Evaluation of Organization Structure

No.	Indicator	Achievement Index*					
		5	4	3	2	1	0
1	Simplicity and speed of decision making						
2	Guidance from seniors to juniors						
3	Efficient crisis management						
4	Data availability for decision making						
5	Empowerment for decision making						
6	Speed of response to market fluctuations						
7	Quality of logistic management between departments, locations, and functions						
8	Interfunctional or interdepartmental cooperation						
9	Overall teamwork of organization						
10	Impact, influence, or dominance of top team on organizational culture						
11	Influence of employee union on organizational culture and decision making						
12	Organization's competency to absorb or tolerate negative results of decisions or events						
13	Systemic support for problem solving						

(Continued)

Table 2.27 (Continued)   Evaluation of Organization Structure

No.	Indicator	Achievement Index*					
		5	4	3	2	1	0
14	Transparency in interdepartmental transactions						
15	Mature processes of leadership building						
16	Independence of junior employees in handling extraordinary situations						
17	Frequency of reporting by juniors to seniors						
18	Volume and quality of reporting by juniors to seniors						
19	Homogeneity in thinking at various employee levels						
20	Simplicity in organizational hierarchy						
21	Timely promotional chances available to employees						
22	Coordinative efficiency between line departments and support departments						
23	Sharing of organization's vision, future plans, and relevant strategies across organization						
24	Facilitation for management of change at various levels						

*5 = Excellent, 4 = Very Good, 3 = Good, 2 = Marginal, 1 = Poor, 0 = Nil.

2. Rank the factors listed in Table 2.28 according to the degree to which they influence the overall quality of your operations management. [Purpose: Understanding factors that determine the quality of operations management.]
3. How do you classify your operations or business activities broadly? [Purpose: Understanding various methods used to classify business operations for various purposes.] See Table 2.29.

Prerequisites of Strategic Planning: Inquiry and Status Evaluation ■ 47

4. What is the overall scope of improvement demonstrated by various areas of your operation in the factors listed in Table 2.30. The rating scale is shown at the bottom of the table. Vertical and horizontal totals should indicate overall scope of improvement on every parameter and in every operation. [Purpose: Assessing the scope of improvement in various parameters by individual operations.]
5. Table 2.31 lists common human factor problems. Based on conditions in your operation, rank the factors by criticality. [Purpose: Critical assessment of employee-related issues that impact operations management.]

Table 2.28   Factors Deciding the Quality of Operations Management

No.	Factor	Rank
1	Organizational structure	
2	Employee productivity	
3	Resource availability	
4	Process logistics	
5	Performance planning	
6	External factors (e.g., market conditions, suppliers, etc.)	

48 ■ *Strategic Planning, Execution, and Measurement*

Table 2.29 Classification of Business Operations

No.	Category	Components			
1	Importance	Core	Support	—	—	—
2	Value creation	High value	Average value	Low value	Zero value	Esteem value
3	Conduct	Insourced	Outsourced	—	—	—
4	Nature 1	Mechanical	Manual	—	—	—
	Nature 2	Complex	Simple	—	—	—
5	Visibility	Tangible	Intangible	Notional	—	—
6	Ownership	Independent	Joint ventures	—	—	—
7	Fiscal support	Subsidized by government	Not subsidized	—	—	—
8	Location	Local	National	International	—	—
9	Interconnectivity	Dependent	Interdependent	—	—	—
10	Salability of output without further processing	Sellable	Not sellable	—	—	—
11	Independent accounting	Possible	Notionally possible	Not possible	—	—
12	Organization competence	Core competence	Subordinate competence	External competence	Dispensable competence	—
13	Capital intensity	High	Average	Poor	—	—
14	Recurring cost	High	Average	Low	—	—
15	Volume	High	Medium	Low	—	—

Table 2.30 Assessing the Scope of Improvement in Operation

Scope of Improvement	Operations*						
	Procurement	Production	Marketing	HR	Finance	Corporate Affairs	General Administration
Quality							
Cost							
Time							
Safety							
Output size							
Flexibility							
Downsizing							

*0 = Nil, 1 = Little, 2 = Reasonable, 3 = More than reasonable, 4 = High, 5 = Very high.

Table 2.31 Employee-Related Issues Impacting Operations Management

No.	Human Factor Problem	Rank by Criticality
1	Impact of union activity	
2	Level of competence	
3	Method of recognition	
4	Employees from local area versus employees from other areas	
5	Emotional gaps between officers and workers	
6	Quality of communication and coordination	
7	Lack of multiskill capability	
8	Duplication of operations	
9	Availability of suitable employees	

Part E: Employee Productivity Inquiry

1. How are your employees rewarded or penalized for good or poor performance? Rank the choices in Table 2.32 in order of preference. [Purpose: Review of performance-linked reward system.]
2. Rank the leadership styles listed in Table 2.33 according to your preference. [Purpose: Appreciation of various styles of leadership.]
3. Rank the leadership attributes listed in Table 2.34 according to your preference. [Purpose: Review of leadership styles.]
4. How do you rank the team-building factors listed in Table 2.35? [Purpose: Comprehensive assessment of various factors to be considered for team building.]
5. Use Table 2.36 to list the five best strengths of your department or division team (based on the factors listed in Table 2.35), then rank them in terms of value in the right-most column. [Purpose: Critical assessment of the strengths of a working team.]
6. List five (or fewer) important weaknesses of your team (referring to the list of factors in Table 2.35) and rank them by the degree of seriousness, in Table 2.37. [Purpose: Critical assessment of weaknesses of a working team.]
7. If you are not on a good team because no team has been formed, note in Table 2.38 five (or fewer) reasons for failure to form a team. Rank the

Table 2.32 Review of the Performance-Linked Reward System

No.	Parameter	Order of Preference
1	Grant or refusal of annual increment	
2	Grant or refusal of periodic promotion	
3	Extraordinary increase or decrease in salary	
4	Change in job content	
5	Change in designation	
6	Geographic transfer	
7	Departmental transfer	
8	Cash reward or penalty	
9	Certificate of merit or remarks of criticism	
10	Symbolic reward or penalty	
11	Other parameter (specify)	

Table 2.33 Styles of Leadership

No.	Leadership Style	Rank by Preference
1	Leadership by power	
2	Leadership by performance and knowledge	
3	Participative or group leadership	
4	Leadership by consensus	
5	Leadership based on situational requirements	
6	Leadership based on job or task requirements	
7	Leadership through ownership	
8	Leadership by age	
9	Leadership based on experience	
10	Leadership based on popularity	
11	Extended or allotted leadership	
12	Other style (specify)	

Table 2.34 Qualities of a Leader

No.	Leadership Attribute	Preference
1	Ability to guide	
2	Ability to make people perform	
3	Ability to manage crises	
4	Capacity to arrange resources	
5	Competence for strategy and policy formulations	
6	Technical competence	
7	Democratic behavior	
8	Respect among followers	
9	Support by seniors	
10	Ability to think ahead	
11	Ability to implement change	
12	Ability to offer or achieve results	
13	Ability to communicate and convince	
14	Appropriate personality	
15	Other quality (specify)	

reasons based on criticality. [Purpose: Critical analysis of the reasons for not forming a good working team.]
8. Assess the leadership attributes listed in Table 2.39 on the 0 to 5 scale shown at the bottom of the table. [Purpose: Relative assessment of your leadership and the leadership of superiors.]
9. How are your four categories of human resources listed in Table 2.40 divided based on degrees of productivity? [Purpose: Statistical analysis of categories of employees based on numbers and costs.]
10. Show in Table 2.41 the contributions of the four categories of employees shown in Table 2.40 to the profit (or economic value added) of the organization? [Purpose: Analysis of contributions made by specific categories of employees to organization profits.]
11. How do you correlate your business divisions with business models and categories of employees? Use Table 2.42 for your analysis. [Purpose: Understanding the correlation between a business model and types of employees.]

Table 2.35 Team-Building Factors

No.	Team-Building Factor	Rank by Importance
1	Clear purpose for building team	
2	Homogeneity in thinking	
3	Loyalty to organization	
4	Loyalty for team leader	
5	Respect for team members	
6	Equal competence of team members	
7	Tenure of team	
8	Knowledge sharing	
9	Sharing of personal problems	
10	Common enemies and hurdles to be faced	
11	Common ambition	
12	Resource availability	
13	Education of team members	
14	Behavioral patterns of team members	
15	Quality of communication	
16	Quantity of communication	
17	Quality of team leadership	
18	Requirements imposed by circumstances or set by top management	
19	Physical closeness	
20	Clear parameters to measure performance of team members	
21	Sharing of team's gains or losses	
22	Team's support to members	
23	Members' readiness to sacrifice for team	
24	System to build consensus within team	
25	Transparency in transactions of team members	
26	Leadership sharing	

(Continued)

Table 2.35 (*Continued*)   Team-Building Factors

No.	Team-Building Factor	Rank by Importance
27	Cultural similarities	
28	Size of team	
29	Ranks or positions of team members in department or organization	
30	Other factor (specify)	

Table 2.36   Strengths of a Working Team

No.	Team Strengths	Rank
1		
2		
3		
4		
5		

Table 2.37   Weaknesses of a Working Team

No.	Assessment of Team Weaknesses	Rank
1		
2		
3		
4		
5		

Table 2.38   Reasons for Not Forming a Good Working Team

No.	Reasons for Incomplete Formation of or Failure to Form a Good Team	Rank
1		
2		
3		
4		
5		

Table 2.39 Comparative Statement of Leadership

No.	Leadership Attribute	Your Leadership (Scale)*	Division Leadership (Scale)*	Organization Leadership (Scale)*
1	Achievement of results			
2	Vision			
3	Respect among team members			
4	Ability to manage change			
5	Ability to raise resources			
6	Ability to build and manage team			
7	Ability to make people work			
8	Ability to reward or penalize team members			
9	Ability to work with other teams			
10	Readiness to sacrifice			
	Total score (of 50)			

*5 = Excellent, 4 = Very Good, 3 = Good, 2 = Marginal, 1 = Poor, 0 = Nil.

Table 2.40 Productivity-Wise Break-Up of Human Resources

No.	Productivity Analysis by Employee Category	Percentage of Total Human Resources	Percentage of Total Cost Represented
1	Systemic		
2	Knowledge based		
3	Decision making (leadership)		
4	Entrepreneurial		

Table 2.41 Employee Category-Wise Contribution to the Profit of the Organization

No.	Employee Category	Percentage Contribution to Profit
1	Systemic	
2	Knowledge based	
3	Decision making (leadership)	
4	Entrepreneurial	

12. How true are the relationships shown in Table 2.43 in your organization? [Purpose: Correlating various categories of employees and types of profits.]
13. Which factors listed in Table 2.44 influence your employees to improve their productivity? Rank the factors for each employee category. [Purpose: Analysis of various factors contributing to employee productivity.]

Table 2.42 Correlation between Business Model and Type of Employees

Business Division	Business Model Dominated by Product, Process, People, Place, or Policy?	Dominant Employee Category: Systemic, Knowledge, Leadership, or Entrepreneurial?
A		
B		
C		
D		

Table 2.43 Category of Employee and the Type of Profit

| No. | Relationship | | Not True | Somewhat True | Considerably True | Totally True |
	Employee Category	Profit Type				
1	Systemic	Economy-based				
2	Knowledge	Industry-based				
3	Leader	Company-based				
4	Entrepreneurial	Distinctive				

Table 2.44 Factors Contributing to Employee Productivity

No.	Employee Category	Factors Influencing Productivity				
		Remuneration	Job Satisfaction	Sharing of Ownership	Job Security	Nonfinancial Recognition
1	Systemic					
2	Knowledge					
3	Leader					
4	Entrepreneurial					

14. How do the categories of employees in your organization (listed in Table 2.45) enjoy (or face) vibrancy of remuneration and job content? [Purpose: Analysis of the impacts of variable pay and job rotation.]
15. What techniques or approaches to improving productivity are preferred by the categories of employees listed in Table 2.46? [Purpose: Understanding of various techniques preferred by different categories of employees to improve their productivity.]
16. How do the categories of employees listed in Table 2.47 respond to organizational failure? Indicate percentages for each response. [Purpose: Critical analysis of employees' responses to organizational failure.]

Table 2.45 Impact of Variable Pay and Job Rotation

No.	Employee Category	Percentage of Remuneration as Variable Pay	Agreement with Change of Job Content			
			Nil	Some	Considerable	Absolute
1	Systemic					
2	Knowledge					
3	Leader					
4	Entrepreneurial					

Table 2.46 Techniques to Improve Employee Productivity

No.	Employee Category	Techniques for Improving Productivity			
		Systematic on-Job Training	External Theoretical Training	Participation in Making and Executing Decisions	Exposure to Peer Performance	Self-Learning
1	Systemic					
2	Knowledge					
3	Leader					
4	Entrepreneurial					

Table 2.47 Response to Organizational Failure

No.	Employee Category	Response to Organizational Failure			
		Fight Failure	Lead Attempts to Fight Failure	Neutral (Individualistic)	Not Bothered (No Action)	Leave Organization
1	Systemic					
2	Knowledge					
3	Leader					
4	Entrepreneurial					

Part F: Resource Management Inquiry

1. How do you rank resource allocations for different business divisions? Table 2.48 shows common factors in resource planning. [Purpose: Understanding importance of factors that determine resource planning.]
2. What weight do you attach to different feasibilities of resource use (Table 2.49) for different business divisions? [Purpose: Critical assessment of feasibilities defining resource use.]
3. Which financial parameters listed in Table 2.50 (with different weights) do you use to determine resource allocations for various projects and/or products? [Purpose: Comprehensive analysis of financial parameters that determine resource allocations.]
4. Which of the resources listed in Table 2.51 are most significant for each business division? [Purpose: Classification of types of assets based on importance and quantities.]

Table 2.48 Resource Planning

No.	Business Division	Determinants of Resource Planning				
		Earning Potential	Resource Availability	Resource Cost	Business Model Risk	Ability to Use the Resource
1	A					
2	B					
3	C					
4	D					

Table 2.49 Resource Usage Feasibilities

No.	Business Division	Dominant Component of Business Model	Feasibility			
			Marketing	Financial	Technical	Sociopolitical
1	A					
2	B					
3	C					
4	D					

Table 2.50 Financial Parameters for Resource Allocation

Business Division	Financial Parameter											
	Absolute Operating Cost	Absolute Cash Operating Profit	Operating Profitability	Operating ROI	Profit after Tax (PAT)	Cash Profit after Tax	Post-Tax ROI	NPV	IRR	EVA (Economic Value Added)	Break-Even Point	Contribution per Employee
A												
B												
C												
D												

Table 2.51 Classification of Resources

No.	Business Division	Dominant Component of Business Model	Resources							
			Physical Long-Term Assets		Intangible Long-Term Assets		Human Assets		Current Assets	
			Rank	% Total Resources	Rank	% Total Resources	Rank	% Total Resources	Rank	% Total Resources
1	A									
2	B									
3	C									
4	D									

Table 2.52 Financing of Resources

No.	Business Division	Financing of Long-Term Assets			Financing of Short-Term Assets (Working Capital)		
		Share Capital	Retained Profits	Long-Term Borrowing	Short-Term Borrowing	Share Capital	Retained Profits	Long-Term Borrowing	Short-Term Borrowing
1	A								
2	B								
3	C								
4	D								

5. How did you finance investments in various long-term and short-term assets? Indicate percentages for the four financing methods cited in Table 2.52. [Purpose: Analysis of various sources of financing.]
6. Rank the factors listed in Table 2.53 that influence your debt/equity mix. [Purpose: Critical assessment of factors that determine capital structure.]

Table 2.53 Factors Deciding Capital Structure

No.	Factor	Rank
1	Tax saving arising from interest payments	
2	Interest rate	
3	Dividend rate	
4	Dilution of promoter's stake through issue of new shares	
5	Increase in earnings per share	
6	Availability of borrowed funds	
7	Stock market conditions	
8	Premium on share issues	
9	Inflation rate	
10	Appetite for future share issues	
11	Appetite for future borrowings	
12	Funds from strategic partner	
13	Government policies	
14	Capacity to serve borrowings	
15	Assets available to increase borrowings by mortgages	

7. Use Table 2.54 to analyze your resource planning. [Purpose: Appreciation of resource planning and review process.]
8. Indicate in Table 2.55 the categories of employees who participate in resource-related matters. [Purpose: Understanding employee participation in resource management at various levels.]
9. Use Table 2.56 to demonstrate how you determine resource requirement at various stages of a product's life cycle. [Purpose: Critical assessment of resource requirements during product life cycles.]
10. Which broad parameters shown in Table 2.57 do you use to evaluate the efficiency of resource utilization? Rank them by importance. [Purpose: Comprehensive understanding of parameters used to evaluate outcomes of resource utilization.]

Table 2.54 Resource Planning & Review Process

No.	Business Division	Resources Identified by Product, People, Process, Place, or Combination	Resource Planning and Review Periodicity		
			Routine Capital Expenditure	Major Capital Expenditure	Working Capital Funding
1	A				
2	B				
3	C				
4	D				

Prerequisites of Strategic Planning: Inquiry and Status Evaluation ■ 65

Table 2.55 Employee Participation in Resource Management

No.	Employee Category	Resource Planning	Monitoring of Resource Consumption	Actual Resource Consumption	Resource Raising	Fund Servicing and Repayment
1	Systemic					
2	Knowledge					
3	Leader					
4	Entrepreneurial					

Table 2.56 Resource Requirement in Product Life Cycle

No.	Approach to Deciding Resource Requirements	Market Study	Product Development	Launch	Take-off	Consolidation	Market Leadership	Revival	Withdrawal
1	Market benchmark								
2	Trend								
3	Earning target								
4	"Gut" feeling								

Table 2.57 Evaluation of Resource Utilization

No.	Parameter	Rank
1	Rate of return on resources IRR & ARR and NPV, PBP identified by ROI, viz: Product Process Place People	
2	Capital (investment) turnover ratio	
3	Incremental capital turnover ratio	
4	Percentage of capacity use	
5	Resource use per employee	
6	Long-term asset turnover ratio	
7	Working capital turnover ratio	
8	Inventory turnover ratio	
9	Receivables turnover ratio	
10	Tangible assets turnover ratio	
11	Intangible assets turnover ratio	
12	Resources invested for every extra degree of quality improvement	
13	Resources invested for every extra degree of productivity improvement	
14	Resources invested for every extra degree of risk reduced	
15	Resources consumed for damage control or recovery	

Prerequisite 2: Status Evaluation

This section assesses 25 parameters in 5 performance areas. After evaluation of the information revealed by the Prerequisite 1 questionnaires, executives should consider growth factors to determine the organization's overall status. The growth factors are categorized as shown below.

Growth Area 1: Business Growth and Expansion
    1.1 Growth in volume
    1.2 Growth in market share

    1.3 Growth through diversification
    1.4 Growth through alteration of value chain (backward, forward, and diagonal integration)
    1.5 Growth through networking (outsourcing, partnering, etc.)
Growth Area 2: Strategic Cost Management
    2.1 Reduction in cost per unit of product
    2.2 Accuracy and effective use of cost measurement and control system
    2.3 Use of cost competence as a parameter for evaluating employees
    2.4 Cost competence as sustained competitive advantage
    2.5 Cost flexibility (discretion to incur costs)
Growth Area 3: Employee Development and Satisfaction
    3.1 Increase in employee productivity
    3.2 Increase in employee satisfaction
    3.3 Productivity as sustained competitive advantage
    3.4 Investment in enhancing productivity (knowledge management, technology improvement, etc.)
    3.5 Productivity initiative as institutional culture
Growth Area 4: Operations, Logistics, and Technology Management
    4.1 Operating profitability
    4.2 Systemic discipline
    4.3 Systemic flexibility
    4.4 Operations improvement
    4.5 Fresh investment in new technology
Growth Area 5: Resource Management
    5.1 Operating Return on Investment (ROI)
    5.2 Owner's ROI
    5.3 Investment turnover
    5.4 Economic value added (EVA) with suitable alterations in computation
    5.5 Liquidity

Versatile Corporate Evaluation

Every executive respondent is responsible for quantifying company performance on each of five parameters using a six-point scale (0 = nil; 1 = poor; 2 = average; 3 = good; 4 = very good; and 5 = excellent) as a component of strategic planning. Note that all the parameters may not be applicable to every company.

Example 2.1—Assume that Mr. A has quantified his company's performance in the right-most columns of Tables 2.58 through 2.62.

68 ■ *Strategic Planning, Execution, and Measurement*

Total score for five areas covered by tables = 13 + 17 + 17 + 21 + 18
= 86 (125) = 69%

Based on Mr. A's perceptions, the composite performance index (CPI) of his company is 69%—between good and very good. In his opinion, the company's performance in the business development area is the lowest (52%) among all five performance areas. Conversely, its performance in the area of 'operations, logistics, and technology management' is 84%—the highest

Table 2.58 Area 1: Business Development

No.	Parameter	Maximum Score	Score Achieved
1.1	Growth in volume	5	3
1.2	Growth in market share	5	2
1.3	Growth through diversification	5	3
1.4	Growth through alteration in value chain (backward, forward, and diagonal integration)	5	2
1.5	Growth through networking (i.e., Outsourcing, Partnering, etc.)	5	3
	Total	25	13
	Score = 52%		

Table 2.59 Area 2: Strategic Cost Management

No.	Parameter	Maximum Score	Score Achieved
2.1	Reduction in cost per unit of product	5	4
2.2	Accuracy and effective use of cost measurement and control system	5	3
2.3	Use of cost competence as parameter for evaluating employees	5	4
2.4	Cost competence as sustained competitive advantage	5	3
2.5	Cost flexibility (discretion to incur costs)	5	3
	Total	25	17
	Score = 68%		

Table 2.60 Area 3: Employee Development and Satisfaction

No.	Parameter	Maximum Score	Score Achieved
3.1	Increase in employee productivity	5	4
3.2	Increase in employee satisfaction	5	2
3.3	Productivity as sustained competitive advantage	5	3
3.4	Investment in enhancing productivity (knowledge management, technology improvement, etc.)	5	4
3.5	Productivity initiative as institutional culture	5	4
	Total	25	17
	Score = 68%		

Table 2.61 Area 4: Operations, Logistics, and Technology Management

No.	Parameter	Maximum Score	Score Achieved
4.1	Operating profitability	5	5
4.2	Systemic discipline	5	4
4.3	Systemic flexibility	5	5
4.4	Operations improvements	5	3
4.5	Fresh investment in new technology	5	4
	Total	25	21
	Score = 84%		

rating among all the performance areas. The variance between the two ratings can have a number of interpretations. For example, the company may be focusing inward instead of seeking opportunities elsewhere, or it may manage its operations very well but fail to pursue a commitment to business development. This could lead to stagnation in the future.

Another possibility is that the weights of the parameters or growth areas may not be the same. In an acute recession, a company may attach more importance to retention of market share and cost control and focus less on diversification, rate of return, and other initiatives. In such situations, relative importances (weights) will have to be assigned to the various parameters (or areas) and a weighted composite performance index computed.

Table 2.62 Area 5: Resource Management

No.	Parameter	Maximum Score	Score Achieved
5.1	Operating ROI	5	3
5.2	Owner's ROI	5	3
5.3	Investment turnover	5	4
5.4	Economic value added (EVA) with suitable alterations in computation	5	4
5.5	Liquidity	5	4
	Total	**25**	**18**
	Score = 72%		

All top executives must meet and participate in "brainstorming" sessions to determine correct scores and weights to be included in a weighted composite performance index. They may not reach a consensus initially. In that case, they should review (and if necessary re-do) their evaluation and attempt to fill the perceptional gaps. They should carry out a second round of brainstorming. If no consensus is reached on the second round, they should consider engaging a consultant to act as a moderator.

The outcome of the Prerequisite 2 exercise will serve as a basis for strategic planning. The results of analyzing the various parameters and performances will be used for SWOT (strengths, weaknesses, opportunities, and threats) analysis of each performance area.

Dimensions of Strength Analysis
Present strengths to be improved further
New strengths to be explored
Present strengths to be retained
Present strengths to be exchanged or combined with new strengths

Dimensions of Weakness Analysis
Present weaknesses to be eliminated or reduced
Potential weaknesses to be avoided or managed
A weakness to be strategically converted into a strength
Averting possibility that present strength will become weakness
Impact of weaknesses on other strengths and vice versa

Dimensions of Opportunity Analysis
Opportunities of volume growth or operational change

Opportunities to be exploited with present and potential strengths
Opportunities that may be lost because of present or potential weaknesses
New opportunities to be created with new strengths
Old opportunities to be replaced by new opportunities

Dimensions of Threat Analysis
Present and potential threats
Opportunities inviting threats
Threats to be converted into opportunities
Old threats to be replaced by new manageable threats
Threats to be averted or reduced by sharing them through networking

After analyzing the SWOT (strengths, weaknesses, opportunities, threats) areas, the results should be verified based on the value drivers of the business as shown in Table 2.63.

Table 2.63 SWOT Analysis

No.	Value Driver	Strengths	Weaknesses	Opportunities	Threats
1	Product				
2	Place				
3	Promotion				
4	Price				
5	People				
6	Processes				
7	Programs				
8	Perceptions				
9	Policies				
10	Projects				

Notes: Place indicates markets and production centers.
Price includes cost performance.
People are all stakeholders (employees, distributors, shareholders, suppliers, lenders, etc.).
Processes are manufacturing, marketing, human resources, and accounting and control functions.
Programs are developmental activities.
Policies include corporate culture and work ethos.
Projects are new investments.

A comprehensive SWOT analysis should facilitate appropriate strategic planning in line with the company's vision and mission. It can also reveal general conclusions that may be useful for further planning of organizational strategy:

Optimistically optimistic (OO): more than 90% of business factors are favorable.
Moderately optimistic (MO): 70 to 90% of business factors are favorable.
Moderately moderate (MM): 60 to 70% of business factors are favorable.
Moderately pessimistic (MP): 50 to 60% of business factors are favorable.
Absolutely pessimistic (AP): more than 50% of business factors are unfavorable.

Generally, an MO scenario is desirable and MM results should be achievable. Therefore, in the strategic plan, two sets of targets need to be decided for these two scenarios.

Chapter 3

Strategic Plan: Components, Preparation, Review, and Recommendations

Structure and Components

A strategic plan is a comprehensive road map for implementing a company's statement of mission so that it can reach the position defined by its vision. The plan consists of a set of strategies for the five growth (performance) areas that collectively define versatile business performance that should enable an enterprise to achieve world-class status. The five growth areas for framing strategies are:

1. Business growth and expansion
2. Strategic cost management
3. Operations, logistics, and technology management
4. Employee development and satisfaction
5. Resource management

The strategies are first defined for each individual growth or performance area and then identified functionally. For example, a sales growth strategy will have to be implemented by sales and marketing but also involve cross functions such as production (for supplies), human resources (for developing suitable competence and/or recruiting sales staff), and finance (for obtaining required working capital).

Duration

A strategic plan may have a duration of 3 to 5 years, depending on organization needs and projection of future events. In a fast-changing global business scenario, enterprises prepare their strategic plans for 3 years. A few enterprises prefer "rolling" strategic plans. A plan is prepared to cover years 1, 2, and 3, then revised after the end of year 1 to cover years 2, 3, and 4. While rolling the strategic plan, experience of year 1 and requirements of year 4 are considered.

Format

A strategic plan starts with five columns that define its five important components as shown in Table 3.1. Each component (or column) of the strategic plan is explained below.

Strategy for Performance Areas

Two alternate strategies should be designed for achieving every target:

Moderately optimistic strategy (MO): workable in an environment in which 70 to 90% of business factors are favorable.
Moderately moderate strategy (MM): workable in an environment in which 60 to 70% of business factors are favorable.

Maximum possible business factors that influence performances in all the five growth areas should be listed and reviewed carefully to determine

Table 3.1 Components of Strategic Plan

Strategy for each performance area (both optimistic and moderate strategies should be noted for each objective or goal)	Hurdle management (describe potential hurdles to be faced in executing each strategy and detail suitable tactics to manage the hurdles)	Cost–benefit analysis (CBA) (detail costs of executing each strategy and explain expected benefits)	Schedule for execution (show start-to-finish schedule covering entire plan)	Crisis management and exit route (describe alternate route or damage control tactic to be followed in case of strategy failure)

which business factors are favorable to the organization. The initial SWOT (strengths, weaknesses, opportunities, threats) analysis and the conclusions based on the strategic inquiry, business model, and value chain of the organization should be reviewed because they provide vital information for compiling a list of favorable factors.

Hurdle Management

Execution of a strategy will be smooth and effective only if the following "hurdle creators" are managed tactically:

1. Competitors and customers
2. Government at all levels
3. Suppliers
4. Distributors
5. Employee unions
6. Financiers including shareholders
7. Environmentalists and environments (sociopolitical and natural)

A set of tactics for dealing with these issues should be defined clearly. The tactics should constitute a proactive approach to hurdle management.

Cost–Benefit Analysis (CBA)

A broad CBA should be performed for every strategy. This exercise should approximately define the capital expenditures and revenue expectations arising from strategy execution. The CBA should quantify the one-time or recurring benefits expected from the strategy over time.

Schedule for Execution

A broad schedule for executing a strategy should be created (with or without brief references to major milestones). The schedule should clearly note starting and ending times of various steps that lead to plan execution. The expected benefit period should be mentioned in this section.

Crisis Management and Exit Route

Some uncertainties may not be obvious while an organization frames a strategic plan. These uncertainties can create crises and ultimately make the strategy ineffective. Damage control and pragmatic exit plans are required components of a strategic plan that projects into the future. Specific exit routes and strategy modifications should be defined along with provisions for handling damages expected from a worst-case scenario.

Execution and Impact Routes

Execution Route

Each strategy is to be executed through specified major milestones that constitute its execution route. The advantages of defining an execution route include:

1. Clear view of each step or milestone on the path to execution
2. Time and resources required for each such milestone
3. Possible hurdles impacting each milestone and suitable tactics to overcome them
4. Exit routes and/or damage control tactics for each milestone
5. Clear method for handling milestones of two different strategies that impact each other
6. Functional involvement at each milestone

Figure 3.1 illustrates an execution route for a strategy planned to expand an organization's customer base in the business growth and development (customer perspective) performance area.

Impact Route

A broad description of the financial (commercial), strategic (long-term), and operational (short-term) impacts of an execution of a strategy should be covered by an impact route. In other words, the impact route should detail the primary and secondary impacts of a strategy on the performance of an organization. Figure 3.2 illustrates the impact route of the strategy execution illustrated in Figure 3.1.

Preparation

A strategic plan should be prepared by the top management of an organization under the leadership of its CEO. The group constitutes a strategic planning team (SPT) and its structure is shown in Figure 3.3.

Strategic Plan: Components, Preparation, Review, and Recommendations ■ 77

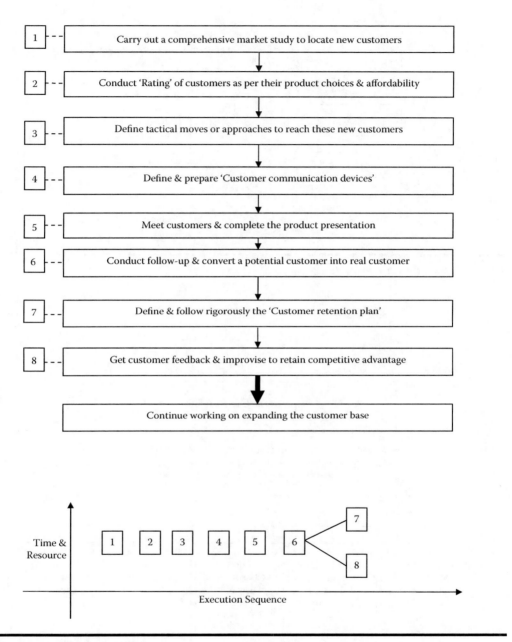

Figure 3.1 Execution route.

The members of all the five SPT subteams and the CEO should work together to frame the strategies over the course of two dedicated meetings held after completion of the preliminary steps of strategic inquiry and status evaluation. They should initially discuss the perceptional and aspirational gaps in their thinking about the present

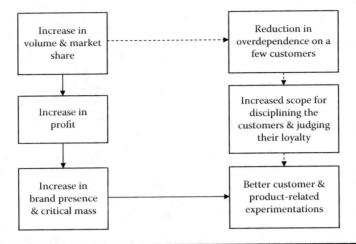

Figure 3.2 Impact route.

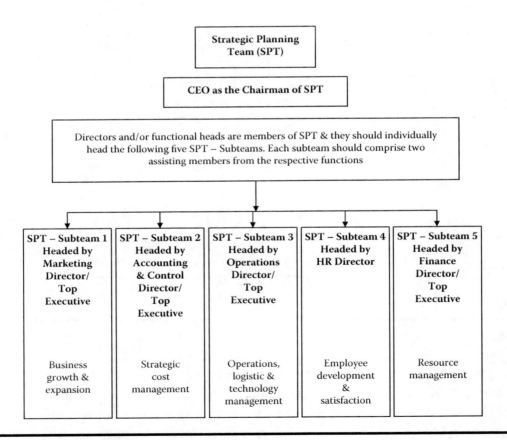

Figure 3.3 Structure of strategic planning team.

status of the organization. This is essential to ensure homogeneity in strategic planning.

The targets of business growth and expansion should serve as important bases for the overall exercise of strategic planning because these targets must be met before an organization can reach the position defined by its statement of vision. Therefore, the strategies for expected business growth and development should become references for the other four performance areas. Of course, the other four performance areas may also act as areas of constraints or limitations for attaining the expected business growth. Hence, the strategies in these four areas must be considered equally critical in deciding the strategies for business growth. In other words, the team must decide whether the hen or the egg is more critical.

In the first major meeting of the subteams (usually planned to last a week), the group should prepare a draft strategic plan under the guidance of the CEO. They should then return to their respective functional divisions and conduct brainstorming sessions to gather ideas for improving strategies cited in the draft plan. The improvisation by brainstorming should be completed in a month, after which all the subteams and the CEO will meet for a second session of 3 or 4 days to finalize the strategic plan.

Implementation

The SPT subteams are responsible for implementation of strategies along with monitoring and measurement of performances in their respective functional areas. A coordination team will be required to deal with interfunctional or interdepartmental projects or activities. For example, a research and development (R&D)-related strategy may require joint execution by members from R&D, marketing, production, and procurement.

The heads of the SPT subteams (who will eventually serve as members of the main team) should meet monthly and discuss the progress of execution of the strategic plan. This discussion can be more effective if monitoring and feedback formats (presented in Chapter 4) are used.

The success of implementation ultimately depends on the quality and execution of an organization's performance management system. The employees at various levels implement the strategies through targets set by management. Therefore, a strategic plan must be converted into

80 ■ *Strategic Planning, Execution, and Measurement*

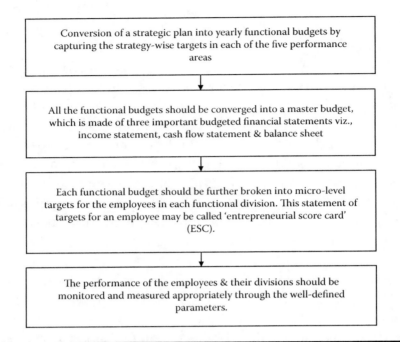

Figure 3.4   Conversion of strategic plan into annual budget.

a set of annual budgets. A budget defines operational and financial targets for each functional division and also for each individual employee. Naturally the effect of a strategic plan depends on how correctly it is translated into annual budgets. Figure 3.4 depicts the conversion of a strategic plan into an annual budget.

Review and Recommendations for Improvisation

A proper structure for reviewing a strategic plan on a quarterly basis is critical. The review process should lead to recommendations for improvisation in executing the plan strategies. The structure for reviewing should be multifunctional, that is, experienced capable managers from all the major functions should be involved in the review and recommendation process.

Five review committees should be formed to cover all five performance areas. Each committee could consist of representatives of all the major functions. For example, a business growth and expansion review committee should include managers from sales and marketing, operations, human resources, and accounting and finance (see Figure 3.5). The other four

Strategic Plan: Components, Preparation, Review, and Recommendations ▪ 81

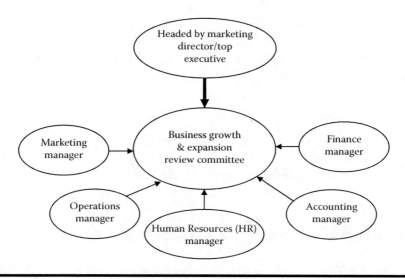

Figure 3.5 Business Growth & Expansion Review Committee.

committees should have the same structure and be chaired by the leaders of the SPT subteams. Reviews should be conducted quarterly and cover the following issues:

1. Outcome of a strategy execution during the immediate past quarter and cumulative results
2. Resources and tactics used for execution
3. Early warning signals of possible strategy failures and plans for damage control or exit
4. Revision of scope, execution route, and schedule of strategy
5. Fresh dimensions impacting the strategy (e.g., uncertainties and risks caused by external environment)

A documented review mechanism presents several important advantages:

1. The ability to make timely changes in a strategy and its execution route improves the overall impact.
2. New ideas should strengthen the execution process.
3. Damage control plans and exit routes should avoid further loss of resources and demoralization after a crisis event.
4. Representatives of various operating functions can provide up-to-date information about the progress of strategy execution in their divisions.

5. New managers participating in reviews are educated about the mechanisms and benefits of strategic management. They also become familiar with operations of other functional divisions. This should help create and enhance organizational homogeneity.

After the review meetings, the functional representatives should return to their respective divisions with a rich knowledge of recent developments, expectations, and fresh ideas. The sharing of knowledge should help in improvising division performance and aid interdivision cooperation.

Generic Strategies for Five Performance Areas

This section provides 150 generic strategies spread over the five major performance areas commonly analyzed as part of the preparation of a strategic plan. Tables 3.2 through 3.6 detail the strategies.

Table 3.2 Performance Area 1: Business Growth and Expansion

No.	Strategy	Possible Benefits	Possible Implementation Hurdles
1	Increase capacities of production and distribution	Higher profit Lower cost per unit Increased reach	Fund availability Ability to manage size Risk of creating idle capacity
2	Increase volume marginally by stretching present capacities	Higher profit No extra cost of investment	Impacts on quality Risk of extra waste and accidents Cost of incentives to employees Higher facility maintenance costs
3	Enter competitor's territory	Unsettle competitors Increase market share Better visibility in new territory Better profit	Competitors' aggressive responses Extra cost of price discounts and credit Entry into price and credit war
4	Create new market	Higher volume Niche market monopoly Geographic diversification	High cost of market penetration Market entry of competitors that incur no costs
5	Develop and launch new product	Product monopoly Distinguished identity Extra volume Risk diversification	Product development and launch costs Product failure Impacts on present products

(Continued)

Table 3.2 (*Continued*) Performance Area 1: Business Growth and Expansion

No.	Strategy	Possible Benefits	Possible Implementation Hurdles
6	Restricted diversification (developing and selling complementary product)	Risk diversification Additional complementary customers Extra profit	Lack of competence Extra investment
7	Total diversification	Diversification of risk New markets Better growth Higher profit Freshness of operations	Scarcity of competence and technology Impact on present business Required funding Possibility of failure Other uncertainties
8	Revival of present product	Product continuity Use of same competence Higher profitability	Additional funds Chance of failure Customer perception
9	Backward integration (backward extension of value chain)	Better stability for value chain Higher value addition to product Higher price and profit Cost reduction Less dependence on vendors	Additional funds Rigidity in design Lack of competence to manage Additional labor problems

10	Forward integration (forward integration of value chain)	Better stability for value chain Higher value addition to product Higher price and profit Cost reduction Less dependence on distributors	Additional funds Rigidity in distribution channel Lack of competence to manage
11	Diagonal integration (portion of semifinished output to be further processed via different value chain)	New product availability Risk diversification Additional profit New market	Additional funds Lack of competence to manage Impact on present product
12	Join competitor as partner and develop oligopoly	Reduction in market risk Reduction in sales overhead Assured volume Better production and sales planning Better price and profit	Competitor agreement Competitor growth leading to future problems Problems under monopoly laws
13	Increase outsourcing	Reduced investment Conversion of fixed cost into variable cost Fewer labor problems Greater growth Better focus on core activity	Quality and supply reliability Cost beyond control Other uncertainties of ancillaries

(Continued)

Table 3.2 (Continued) Performance Area 1: Business Growth and Expansion

No.	Strategy	Possible Benefits	Possible Implementation Hurdles
14	Develop networks with competitors for specific purposes (e.g., marketing, processing, liaison, sourcing, etc.)	Cost reduction Better bargaining power with suppliers Interdependence leading to maturity among competitors	Failure to get competitor's agreement Difficulty of retaining identity Possible difficult exit from arrangement
15	Lease out idle capacities	Better use of capacities Additional revenue Exploitation of competitor's dependence on our facilities Better absorption of fixed costs	Inability to reduce idle capacity permanently Competitor misuse of idle capacity Threats to information confidentiality Impact on present business
16	Use idle capacities of competitor	Reduction of investment Improved relations with competitors Reductions of fixed costs Understanding of strengths and weaknesses of competitors	Dangers of depending on competitors Exposure of competitor to product performance data Risks of exposure of key employees and key data Product distinction may be lost

17	Develop low value–low price products for low end customers	Bigger market share Extra profit Better market visibility Risk diversification Better facility use and recovery of fixed costs	Image problem Lack of competence to manage Competition from unorganized sector Impacts on high value products
18	Develop high value–high price products for high end customers	Additional market Better profit realization per unit Potential for expansion Better use of facilities	Lack of competence to manage Additional investment Impacts on present products and customers
19	Create entry barriers for new competitors	Retention of market share Avoiding future cost of managing competition Maintaining perceptions of existing customers Improved concentration and exploitation of present market	"Boomerang" result Costs Need for political management and possibly undesirable political partners Lack of sustainability
20	Promote (or advertise)	Brand equity Customer retention Local appeal Employee satisfaction Market leadership	Cost Requirement for greater customer commitment Promotion may fail

Table 3.3 Performance Area 2: Strategic Cost Management

No.	Strategy	Possible Benefits	Possible Implementation Hurdles
21	Achieve economy of scale through greater volume	Reduction in cost per unit Better prices to customers and more sales	Risk of unsold stock Uncontrollable increase in volume-based incentive Idle capacity during recession
22	Outsource from less costly suppliers and ancillaries	Considerable cost saving Less investment and higher ROI	Uncertainty of supply and quality Unpredictable factors to add to cost Increase in logistic related cost
23	Develop strategic relationships with suppliers	Long-term cost predictability and control Reliable supplies Better quality control	Working with only one supplier reduce cost flexibility Limited supply sources
24	Invest in suppliers' businesses	Better cost negotiation Control of quality and quantum of supplies Possible change in value chain	Additional funds required Recovery of funds Suppliers readiness to accept arrangement
25	Automation of processes	Reduced labor cost Better accuracy Fewer labor problems Higher production rate	Additional investment Obsolescence of technology Maintenance cost Labor laws

26	Operational logistics improvements	Reduction in operational overhead Fewer delays	Suitable competence required at all levels External factors beyond control
27	Convert maximum portions of fixed salaries into performance-based (variable) salaries	Reduction in committed cost Better employee productivity	Unions agreement Upper cap on variable pay to be strategically decided
28	Replace experienced employees with less experienced employees	Reduced salary costs Increases in volume and speed	Labor laws Cost of separation Cost of training of new employees Loss of expertise
29	Share common facilities with sister companies	Cost sharing Shared maintenance responsibility Investment sharing	Difficulties of emergency management Risk of confusion in logistic arrangement
30	Develop production centers in poor countries	Low production cost Better cost control due to predictability	Sociopolitical problems Difficult exit Poor infrastructure
31	Share infrastructure with other local businesses	Cost sharing Investment sharing Sharing of future risks Better management of government issues	Cooperation problems Logistic confusion

(Continued)

Table 3.3 (Continued) Performance Area 2: Strategic Cost Management

No.	Strategy	Possible Benefits	Possible Implementation Hurdles
32	Develop skills to manage government issues	Reduced delay costs Reduced infrastructure costs Reduced direct and indirect taxes Better subsidies	Lack of required competence Ethical dilemmas Political issues Cultural issues
33	Develop relationships with local politicians	Avoidance of costs of handling labor unrest Local infrastructure at reasonable price Better management of sociopolitical elements	Political interference Decline of business concentration Competitor use of same tactics
34	Standardize products and processes	Improved cost measurement and control Promotion of cost reduction efforts via benchmarks	Standardization may cause rigidity Product distinction may not be promoted
35	Arrange collective sourcing of materials and services with competitors	Better bargaining with suppliers Better control of supply chain Sharing of logistics costs	Competitors may not cooperate Product distinction may not be possible
36	Join competitors to negotiate with unions and governments	Better bargaining to control wage hikes Lobbying to reduce taxes and increase subsidies	Competitors may not cooperate Risk of cartel mechanism and legal consequences

37	Exploit brand equity or extend mother brand i.e., primary brand	Reduced promotional costs Better customer confidence Better visibility	Extension of brand may boomerang Brand-based pricing may be perceived as high
38	Develop accurate cost recognition, measurement, and control system	Better cost control Cost reduction becomes visible Better product pricing Better process-related decisions	Common costs and assets shared by two products are difficult to separate Cost data transparency may be dangerous
39	Use cost control parameters in employee performance appraisal	Better cost awareness Increased cost reduction effort Promotion of entrepreneurship	Extreme use may cause accounting manipulations Developmental costs may be postponed Cost reduction may be achieved by diluting quality and safety
40	Use concept of life-cycle costing	Better correlation of product utility and recurring cost Better accounting for incremental capital costs	Difficult predictability of life-cycle costs Uncertainties of life-cycle performance
41	Study and advise suppliers and distributors to reduce their costs	Cost savings Better ROIs for suppliers and distributors	Transparency and reliability of cost data Legal complications

(Continued)

Table 3.3 (Continued) Performance Area 2: Strategic Cost Management

No.	Strategy	Possible Benefits	Possible Implementation Hurdles
42	Conduct thorough market research	Competitors' cost efficiency could be benchmarked Costs of product launch and promotion can be channeled appropriately	Cost of market research Authenticity of research findings Periodicity of research
43	Inculcate entrepreneurship through sharing of ownership (stock plans, profit sharing, more empowerment)	Enhanced employee awareness of cost reduction efforts	Employee cooperation Owners' view of sharing ownership with employees
44	Application of concept of opportunity cost	Employee awareness of market-driven costs Sharper cost efficiencies Better entrepreneurship	Extreme use of concept is unrealistic and would demoralize employees Cost may not be viewed as competitive advantage
45	Make organizational structure nimble and clear	Avoidance of delays could reduce time-related costs Improved feedback about cost incurrence	Orthodox legacies are difficult to change Owners' priorities Union resistance Legal hurdles
46	Add value to product for better price realization	Better alignment of cost and price Readiness of customers to accept extra cost	Lack of competence Customer affordability Customer perceptions Impact on low value products

47	Develop market-related predictability for strategic purchases	Price increases may be projected Utilize advance purchases to save inflationary costs	Lack of competence High cost of carrying inventory
48	Import material from countries whose currencies are weaker than home currency	Exchange ratio advantage Less dependence on domestic suppliers Global exposure	Unfavorable exchange ratio Questionable reliability of foreign suppliers International legal and process complications
49	Replace costly loans	Reduction in interest cost	Legal replacement may be difficult Requirement for negotiating skills
50	Improve planning skills across organization	Better planning to reduce cost of delays Better clarity for crisis management	Excessive planning may create bureaucracy Lack of competence at lower levels
51	Apply real or notional profit center concept to all functions	Market-based benchmarking Vibrant entrepreneurship Visible impact of cost control on profitability Sense of ownership	Real or notional application may not be possible due to inadequate data Extreme application may demoralize employees

Table 3.4 Performance Area 3: Operations, Logistics, and Technology Management

No.	Strategy	Possible Benefits	Possible Implementation Hurdles
52	Design simple but comprehensive operations control parameters at all organization levels	Visibility of control systems Better control and better results	Lack of skill for designing parameters Acceptance at various levels
53	Locate and outsource operations strategically	Cost reduction Better quality Better choices of suppliers and products Less risk of obsolescence	Dependence on outsiders that may not be reliable Future joint venture partners may not be comfortable with excessive outsourcing Bigger competitors with less outsourcing may defeat the model
54	Develop operational catalogs for self-management	Automatic training of workers Better crisis management Less pressure on managers Improved output and organizational culture	Lack of competence to develop Acceptance by workers Piracy of catalogs by competitors
55	Develop and inculcate operational benchmarking as organizational culture	Continuous improvement in performance Better exposure to market realities Reduction in cost	Source of benchmarks Reliability of benchmarks Acceptance by workers

56	Correct organizational structure to improve operational logistics	Better operational flow Better efficiency and cost savings	Orthodox thinking Lack of ability to alter Worker responses Cost factors
57	Join competitors to create better infrastructure (roads, rails, marine transport facilities)	Better movement of goods and people Higher earnings and customer satisfaction Division of responsibility to raise funds and operate infrastructure	Competitors may not agree for collaboration May strengthen competitors Government policies Required funding
58	Develop strategic sourcing and vendor management system	Reliable supply Reduced cost Better prediction of supply prices and quality	Vendors may not agree Rigidity in sourcing
59	Develop and use a wide distribution channel	Better product visibility and availability Better market penetration Better crisis management Improved customer feedback	High investment cost Sharing margins over many layers is difficult

(Continued)

Table 3.4 (Continued) Performance Area 3: Operations, Logistics, and Technology Management

No.	Strategy	Possible Benefits	Possible Implementation Hurdles
60	Improve worker participation in quality and cost initiatives	Better product quality Reduced cost Better employee satisfaction Improved exposure to higher quality benchmarks Improved efforts of development across organization	Excessive transparency may be dangerous Lack of worker competence Scope and suitability for participation
61	Create a matrix structure for reporting and guidance	Guidance and monitoring of every employee by a functional superior and location leader	Worker confusion Increase of structural complexity
62	Locate production centers in undeveloped areas or countries	Tax holidays Lower salaries and other input costs Better government support Better employee loyalty	Sociopolitical problems Low employee productivity Inferior infrastructure Uncertain governmental policies
63	Develop and use multipurpose or flexible technology where possible	Reduced capital and recurring costs Less technological obsolescence	Difficulty in obtaining flexible technology at reasonable cost Technology may not fit employee productivity and other operational logistics Possible product quality dilution

64	Develop direct marketing channel	Better margin per unit Direct customer contact Better understanding and control of market Improved brand equity	Larger investment Lack of direct marketing competence Time pressures on entrepreneurs Direct exposure to market risk
65	Develop strong brands	Increased sales and profits Customer satisfaction Better competitive position Scope for brand extension to other products Reduction of promotional cost over long run	Heavy investment No guarantee of brand earning power Notional asset Uncertain realizable value Branded costs
66	Devise a product-wise organizational structure	Clear view of product-wise performance Better integration of functions Accuracy in product costing and pricing	Dilution of functional control Demoralization of functional heads
67	Devise a functional structure of the organization	Improved functional concentration Improved functional cost recognition and reduction	Organizational split leading to bureaucracy Difficulty of product costing and lack of visibility

(Continued)

Table 3.4 (Continued) Performance Area 3: Operations, Logistics, and Technology Management

No.	Strategy	Possible Benefits	Possible Implementation Hurdles
68	Develop and use cooperative arrangements with academic and technical organizations	Improvement in technological research leading to improvisation Better knowledge management practice Sourcing good technocrats	Readiness of technical institutes to collaborate Possibility of excessive theoretical work
69	Develop small operational activity groups consisting of mixes of workers	Versatile understanding of business Channel for worker suggestions and grievances Platform for leadership development	Time-consuming project Union responses Potential for gossip and political activities
70	Educate all employees about the whole value chain	Improvement of employee performance Improved interfunctional feedback and collaboration	Employee inability to appreciate value chain Potential for excess transparency
71	Expose workers to higher operational benchmarks gathered from various sources	Better benchmarking leading to better performance Better wage management	Source of benchmarks Worker responses Frustration caused by very high benchmarks

72	Create small enterprises to cultivate future value chain expansions	Tapping future value chain expansion via organic approach Useful exposure for key employees Ability to create growth-oriented roles for employees Better financial and organizational management	Unsuccessful enterprises may have to be closed and impact bottom line. Few employees may be willing to work in these futuristic enterprises
73	Train workers to understand the costs of their operations	Better cost awareness to achieve better cost management Improved maturity would reduce undue demands	Lack of worker competence and little desire to understand costs Excess cost transparency may be dangerous
74	Develop unconventional energy sources and partners	Reduced energy costs Uninterrupted energy supplies over long run Fewer environmental problems Possible competitive advantage	Lack of ability to develop sources and partners Capital and recurring costs
75	Develop short-term strategic alliances with technologically advanced organizations	Better technology and working systems Exposure to large-scale operations Advancement at lower cost	Domination by larger organizations Contract may not allow easy separation or independent use of new technology Overdependence may dilute in-house capabilities
76	Connect operational benchmarks with financial benchmarks of performance	Better appreciation of technical and financial results Improved use of resources	Equivalence may be difficult Potential for excess transparency

(Continued)

Table 3.4 (Continued) Performance Area 3: Operations, Logistics, and Technology Management

No.	Strategy	Possible Benefits	Possible Implementation Hurdles
77	Differentiate operating and non-operating assets	Better asset monitoring Accurate computation of financial and operating indicators Effective use or disposal of non-operating assets	Differentiation may be difficult Non-operating assets may be strategically important in the long run
78	Develop operational team leaders as counselors	Improved employee relations through shop floor counseling Better environment and productivity	Lack of ability of senior employees to become counselors Employee responses Ineffective counseling is counterproductive
79	Combine new and experienced employees in all operational teams	Good mix of energy and experience Better training of new employees More enthusiastic experienced employees	Acceptance by experienced employees Attention required by new employees
80	Design and use visible operational benchmarks to determine penalties and rewards	Better performance management Improved transparency	May be counterproductive Tactical flexibility may be lost through excess transparency

81	Convert functional or operational departments into separate legal companies	Focused performance management Avoiding need to allocate common costs Better transfer pricing through visible and real benchmarks	Increased administrative costs Loss of synergistic advantages of common management and facilities Impacts on tax planning
82	Connect all operating departments through intranet to allow strategic use of business information	Improved online feedback and control Faster accounting results Better benchmarking	Costly installation Lack of competence to develop system Potential for excess transparency
83	Centralize all purchases for all divisions	Better coordination Increased volume should create bargaining power to lower costs and improve quality	May not be logistically suitable Redistribution may be time consuming and costly
84	Develop a system of locating non-value adding or low-value adding operations	Reduction of operating and capital costs Better focus on value adding operations Strengthening core competence	Employees need extra motivation Separating or deleting low value adding operations may be difficult

(Continued)

Table 3.4 (Continued) Performance Area 3: Operations, Logistics, and Technology Management

No.	Strategy	Possible Benefits	Possible Implementation Hurdles
85	Compel every employee to participate in research and development efforts	Faster operational development Better employee satisfaction Improved research culture	Employees may not be able or willing to participate Requirement may be counterproductive Overall coordination may be costly and difficult Possible negative impacts on routine work
86	Ensure that senior managers set examples for workers	Worker motivation Better management relations with workers Better industrial relations	Implementation may be difficult Excessive closeness may be undesirable Employees may construe activity as a "gimmick"
87	Develop a high-tech operational system for future installation and train new employees for future tasks	Better strength to compete in future Improved morale of new employees Better operational and investment planning	Effect on present work Cost factor Limitations on advance planning

88	Implement flexible timing system for top and middle management	Improved work culture and morale Better employee retention	Administration difficulty Potential for misuse Lower management staff and workers may want the system
89	Develop employee housing colony near factory	Improved attendance Better management of overtime work Better family culture	Investment in housing Office rivalry may extend to housing Employees may prefer other housing Potential liability
90	Develop school, medical, and other facilities for employees	Better organizational culture Improved employee attendance Cost savings Better quality of life for employees	Fund availability Administration burdens Employee acceptance Lack of commercial viability

Table 3.5 Performance Area 4: Employee Development and Satisfaction

No.	Strategy	Possible Benefits	Possible Implementation Hurdles
91	Define versatile productivity for all (connect it to cost, time, safety, and quality)	Versatile performance at every level Clear understanding of business requirements	Versatile definition may be subjective Labor union may not agree Implementation may be difficult
92	Develop productivity-based remuneration	Direct increase in productivity Better employee maturity Better cost control and higher profit	Labor resistance Connection of productivity and pay may be subjective Labor union may insist on incentives despite recession conditions
93	Empower managers to enhance their productivity	Better results More ability to manage crises Improved organization loyalty Improved leadership	Empowerment without control may be risky Degree of empowerment is subjective decision Empowerment may not follow systems
94	Ensure organization's structure, group, and individual roles are clear	Improved productivity Fewer bureaucratic hurdles	Resistance of some workers Subjectivity in role definitions
95	Create clear and complete working systems	Smooth system operation Fewer errors Self-guidance by employees Better work atmosphere Fair wages Scope to improve	Some assignments may be subjective Possible increases of rigidity and bureaucracy Standards may become outdated if not reviewed and modified regularly

Strategic Plan: Components, Preparation, Review, and Recommendations ■ 105

96	Create excellent working conditions	Improved productivity Better employee relations Cost reduction Better scope for bringing change Better crisis management	Investment required Maintenance of working conditions
97	Support employees through parallel or innovative methods of career building	Long-term employee sustainability Better productivity Culture of innovation	Costly and time consuming Employee expectations may increase
98	Provide employees with better technology and raw materials	Productivity increase Improved product quality Cost and profit advantages Better worker relations	Large investment Greater recurring costs Lack of skills to manage better technology
99	Handle labor unions strategically and tactically	Emphasis on productivity-based wages Higher productivity benchmarks	Lack of knowledge of union relations External political interference
100	Use innovative methods of empowerment, succession planning, and leadership grooming	Better employee satisfaction and performance Orderly succession Ability to handle greater challenges	Risk and expense Possibilities of unequal opportunities Possible increase of employee expectations

(Continued)

Table 3.5 (Continued) Performance Area 4: Employee Development and Satisfaction

No.	Strategy	Possible Benefits	Possible Implementation Hurdles
101	Improve planning of production, sales, purchases, logistics, and factory layout	Better use of productivity Reductions in costs and waste Better response to market conditions Improved supply chain management	Lack of scope to improve Lack of competence to improve Requirement for additional resources Employee responses
102	Improve movements of employees, equipment, and other assets	Time and resource savings Better productivity and output Improved employee satisfaction Improved working conditions and asset life	Inability to achieve changes Employee responses Nature of assets
103	Distribute details of daily and weekly targets with notes on self-guidance	Improved productivity More employee initiative Management by exception	Lack of employee support Lack of competence at all levels Cost and time required to devise self-guiding charts Continuity in efforts
104	Expose all managers and executives to strategic thinking and commercial acumen	Better and sustainable results Focus on end results Gradual development of leadership mind-set	Difficult, time-consuming, and expensive implementation Increase of employees' role-related ambitions

105	Inculcate and improve productivity maintenance	Timely maintenance at lower cost Minimal production losses Improved equipment handling	Union resistance Lack of training facilities Increased use of equipment
106	Create internal union and promote positive leadership	Elimination of external interference Better manager–worker relationships Better employee morale Informality expedites decision making	Hindrance from external sources Worker immaturity Lack of HR competence
107	Convert qualitative parameters into quantitative benchmarks	Better objectivity in performance appraisal Easy link of productivity and incentives Better work atmosphere	Subjectivity in basic quantification Lack of union cooperation
108	Develop and use effective knowledge management system	Better, faster, and more economical development of human resources Turnover reduction Better organizational culture	Lack of ability to develop system Lack of union cooperation Complex structure of organization Reluctance of owners to provide facilities

(Continued)

Table 3.5 (*Continued*) Performance Area 4: Employee Development and Satisfaction

No.	Strategy	Possible Benefits	Possible Implementation Hurdles
109	Connect business plan to competencies required and training to be offered	Relevant focused training Effective resource utilization Better employee productivity and organizational culture	Unclear business plan Lack of a system to study competence gaps
110	Encourage excellent employees to propose role changes or pursue greater responsibilities	Channel for employees to express their capabilities Employees can study value chain to identify opportunities for growth Voluntary activities allow employees to show readiness to handle greater challenges	Unjustified increase in employee expectations Unequal employee opportunities Less aggressive employees may be viewed as not capable
111	Expose key employees to best work practices of other organizations	Enhanced motivation Productivity improvements Better benchmarking	Lack of union cooperation High cost of program Availability of programs allowing such exposure
112	Involve employees' spouses in family and community activities	Greater employee loyalty Better productivity	Reaching families may be difficult Union may consider approach as a threat
113	Develop appropriate career plans for new employees	Improved employee retention and productivity Better human resource planning	Complex organization structure Lack of response from employees Competence of HR executives

Strategic Plan: Components, Preparation, Review, and Recommendations ■ 109

114	Develop comprehensive and acceptable retirement benefits	Guarantee of postretirement income Increased employees confidence and loyalty	Lack of ability to develop program Communication gap between employees and management Inadequacy of funds Legal complexity
115	Develop automatic system to monitor job attendance	Better supervision Improved productivity	Union opposition Employees misunderstanding Impact on assignment flexibility
116	Simplify organizational structure	Better and faster decision making Improvement in organization culture Better scope for change management	Opposition by ambitious executives Satisfaction with existing structure Psychological impacts
117	Design employee-friendly salary structure	Employees' specific needs are met Better tax and savings planning	Rigidity of tax structure Difficulties of administering program Employee expectations
118	Offer consultant's role to employees to allow better tax planning	Reductions in employee income taxes Savings on long-term and retirement liabilities	Rigidity of tax structure Employees responses Reduced accountability of consultant

(Continued)

Table 3.5 (Continued) Performance Area 4: Employee Development and Satisfaction

No.	Strategy	Possible Benefits	Possible Implementation Hurdles
119	Reduce overtime work culture	Reduced wage costs Elimination of go-slow mentality	Employee responses Difficulty of meeting goals within time limits
120	Increase employee participation in target-setting activities	Improved employee commitment leading to greater achievements Setting of realistic and acceptable targets	Time consuming Lack of employee ability to determine targets Insecurities of managers and owners may impact acceptance
121	Recognize and reward outstanding employees outside the regular system	Improved retention of outstanding employees Motivation of others to become outstanding No impact on existing salary structure	Union opposition Impact on organization's culture if unrewarded employees feel humiliated
122	Devise specific salary plans for different locations, plants, products, etc.	Better conformity to regional differences in costs of living Ability to base salaries on location, plant, or product earnings	Union opposition Demoralization of lower paid workers Tendency of employees to choose only locations allowing high earnings

123	Pay more for postretirement days and less at present	Improved employee retention Compulsory savings for postretirement use Reduction of income taxes	Acceptability of present low salary More tax on retirement benefits Impact on company tax planning
124	Pay consolidated salary without defining allowances, benefits, basic rates, etc.	Reduction in accounting cost Flexibility Better tax planning	Employees may not understand system Tax regulations may hinder employee tax planning Employee classification may be unclear
125	Clearly differentiate roles of owners and managers	Improved manager freedom and accountability Owners can pursue other interests Periodic reporting system would allow owners to have better control	Owner insecurity Improper control system Employee inability to use more freedom

112 ■ *Strategic Planning, Execution, and Measurement*

Table 3.6 Performance Area 5: Resource Management

No.	Strategy	Possible Benefits	Possible Implementation Hurdles
126	Conduct business with own funds as much as possible	Low interest cost Low risk of debt service Less interference by lenders	Loss of tax savings and low interest rates Owners will lose the advantages of 'trading on equity' Limitations on fund availability
127	Conduct business with borrowed funds as much as possible	Maximum tax advantage No dilution in ownership Inflationary benefit No perpetual liability for serving equity	High risk caused by interest and loan burden Possible liquidity problem Difficulty of predicting interest rates
128	Use foreign loans at lower interest	Low interest cost Possible advantage from exchange rate Less interference by foreign lenders	Risk of high volatility in exchange rates Strict lender conditions Uncertainty caused by international events
129	Funding of new projects should be gradual as a project progresses	No cumulative burden of high interest No idle funds Reduced fund waste	Availability of funds when needed Possible increased interest rate on progressive disbursement of loan Difficulty of predicting need
130	Charge interest on use of own funds	Users of funds are disciplined Operating earnings are better monitored Owners feel rewarded	Difficulty of benchmarking notional interest rate Lack of tax advantage on notional interest Owners should only expect profit & not interest

131	Acquire assets to ensure future strategic advantage	Strategic advantage Price advantage May prevent entry of new competitors	Funding required Possibility of idle investment
132	List company on stock market	Better visibility Share price indicates company performance Vehicle for fund-raising	Stock market price may be illusory Extra visibility may lead to takeover threat
133	Always maintain high liquidity	Eliminates liquidity crises Allows exploitation of unexpected investment opportunities Cash purchasing provides edge over competitors Intragroup lending is possible	Control issues High liquidity is costly Funds are idle Funds may be misused Bank funds are available whenever required
134	Develop strategic relationship with bankers	Facilitates all-season cash management Low interest rates Edge over competitors	Flexibility of dealing with multiple banks is lost Bank failures may become company problems
135	Create hedge with a foreign country by combining imports and exports	Avoids risks of volatility in foreign exchange rates Better commercial and financial equilibrium	Hedges may not be possible Restricting activity to one country may be dangerous

(Continued)

Table 3.6 (Continued) Performance Area 5: Resource Management

No.	Strategy	Possible Benefits	Possible Implementation Hurdles
136	Conduct business with negative or zero working capital	Lower cost of capital Improved ROI Elimination of dead inventory and bad receivables	May not be possible May be dangerous during critical times May be achieved at the cost of business development
137	Involve business units in raising and managing funds	Better fund management Shared accountability More financial executive time for strategic activities	Lack of competence Difficulty of coordination
138	Outsource fund management, internal audit, cost audit, and other functions	More focus on value-adding finance functions Reduced cost of support functions	Loss of confidentiality Possible lack of competence of contractors Reduced in-house entrepreneurship Difficulty of coordination
139	Require all business or functional divisions to prepare financial feasibility analysis reports of all investment proposals	More financial awareness improves fund use More financial executive time for strategic activities Faster decision making	Lack of ability to perform analysis Reduction of confidentiality of financial information

140	Issue shares to small investors throughout country	Larger investor base Fewer threats to control	Increased administrative cost Insufficient numbers of small investors During difficult times, small shareholders may not unite to protect their interest & suffer a lot
141	Expand business by financing with equal proportion of company's own funds and borrowed funds	Good and balanced debt/equity ratio Improved confidence of shareholders and lenders High debt may provide tax benefit	Company may not always have required funds for expansion
142	Securitize all trade receivables to raise funds (factoring)	Easy fund availability Ability of marketing executives to concentrate on selling instead of recovery	Complexity of process High cost Risk of dealing with customers with unstable creditworthiness
143	Create separate company to manage group's finances	Improved management of group finances Lower cost of funds	Centralized fund management may not be possible Participating companies lose flexibility
144	Use funds of venture financiers for risky projects	Transfer of financial risk to venture financiers Ability of technocrats to focus on business	High cost of venture financing Strict conditions for conversion and repayment
145	Assign high weight to fund usage efficiency in performance evaluations	Improved fund use Better allocation of funds	Neglect of other performance factors Excessive use of technique may be frustrating

(Continued)

Table 3.6 (Continued) Performance Area 5: Resource Management

No.	Strategy	Possible Benefits	Possible Implementation Hurdles
146	Utilize more leased assets and fewer company owned assets	Increased ROI No risk of obsolescence Replacement flexibility	High rents Decreased balance sheet strength Low mortgaging capacity No guarantee of availability of new leased assets Lease agreement may limit use of assets
147	Allocate funds for projects based only on IRR (internal rate of return)	Better financial feasibility and discipline Productive use of funds Elimination of risky ideas	Neglect of other important projects Difficulties of benchmarking IRR Frustration of technocrats
148	Self-funding of every new project	Fewer funding requirements Better culture of entrepreneurship Faster expansion of organization Better rewards to performers	May not be always possible Frustration of technocrats Restricted choices projects

149	Resources managed by owners, not by executives	Better and objective use of funds Freedom of executives from resource management Better accountability of owners Focused fund-raising and deployment	Lack of competence of owners Failure of executives to use funds optimally Owner business philosophy may adversely affect performance Increased owner dominance may frustrate executives
150	Resource management to be based on single parameter of economic value added (EVA)	Reasonably assesses efficiency of fund use Attaches notional cost to reserve funds Distinguishes fund suppliers and fund users (owners and managers)	Benchmarking of cost of capital is difficult Requires allocation of common costs and common facilities Fails to properly consider strategically acquired assets and assets under development

Chapter 4

Monitoring Parameters for Strategy Execution and Impact Measurement

Need for Monitoring

A very good strategy may prove to be wrong if not executed well. Every milestone of the execution process must be monitored (supervised) systematically. Monitoring of the execution process will reveal early warning signals of problems and allow management to measure the related impacts. Monitoring serves many purposes:

1. It ensures appropriate (step-by-step) implementation of a strategy.
2. It facilitates instant evaluation of each completed phase of the execution process; weaknesses in execution quickly become visible and timely corrective actions can be taken.
3. Monitoring naturally includes reasonable attempts to measure positive and negative impacts of strategy execution. Such measurement may be quantitative or qualitative or both.
4. The early warning signals revealed by monitoring define the need for strategy revision or replacement. The revision may be in the tactical approach or relate to resources used or scheduling.

5. Monitoring also facilitates the performance measurement of functional and collaborating divisions working on a strategy. Such performance measurement may be reduced to the micro level for evaluating individual employees through their entrepreneurial score cards (ESCs).

Periodicity of Monitoring

The periodicity of monitoring may not be the same for every strategy because the impacts of strategy execution may be qualitative or intangible and also long term. A few strategies may be executed in a short time and thus require very alert monitoring. Their impacts may be short-lived. It is normally difficult to decide exact periodicity of monitoring for certain types of strategies:

1. Interdivisional strategies, e.g., product development
2. Complex or multifaceted impacts arising from large projects, e.g., capacity expansion
3. Strategies related to external environment and surrounded by too many uncertainties, e.g., attempts to obtain certain fiscal benefits from government agencies
4. Strategies involving long-term qualitative impacts, e.g., fundamental research intended to lead to a new product or formulation
5. Strategies for changing old rigid policies, e.g., deciding against using local agents for product promotion in small markets
6. Strategies of a highly sensitive nature, e.g., introducing a performance-based variable pay plan
7. Strategies of an abstract nature, e.g., changes in organizational structure

Monitoring Process

Performance monitoring statements should be prepared by the respective strategic planning team (SPT) subteams for their performance areas. The leaders of the subteams should submit the statements to the SPT. The SPT members should discuss the statements thoroughly and also review the feedback and recommendations made by the review committees. The SPT should determine changes in strategies or new strategies to be framed to achieve the defined results.

The parameters to be used for monitoring should be selected objectively. They should exactly represent the purpose of the strategy so that they can capture correctly the positive and negative impacts of executing the strategy. The monitoring of strategy execution and measurement of the impacts should be performed by comparing actual and expected results. As noted earlier, the expected results may be moderately optimistic (MO) and/or moderately moderate (MM). All differences or deviations should be analyzed and the reasons determined so that further action can be taken.

Illustration 4.1—Table 4.1 illustrates a monitoring exercise. The exercise defines two parameters based on the purpose of the strategy: (1) sales growth through market penetration (parameter 1) and the ancillary purpose (2) is building an infrastructure of new dealers (parameter 2). The performance area monitored is business growth and expansion. The strategy monitored is recruiting of new dealers for neglected markets; monitoring periodicity is quarterly.

Table 4.1 Quarterly Monitoring Statement

No.	Monitoring Parameter	Optimistic Expectation (a)	Moderate Expectation (b)	Actual Performance (c)	Deviation 1 (a) – (c)	Deviation 2 (b) – (c)	Reasons for Deviation
1	Sales growth based on recruitment of new dealers in neglected markets	$5 million	$4 million	$4.2 million	$0.8 million (adverse)	$0.2 million (favorable)	Neglected markets not fully penetrated; quarter period inadequate for market penetration
2	Readiness of infrastructure of new dealers	Nil	Sales office and warehouse to be ready by next quarter	Both ready	Nil	Zero	No deviation; complete performance

Monitoring Parameters

Tables 4.2 through 4.6 list the specific performance statements and their typical reporting periods used to monitor company performance. The statements are classified into the five standard business areas discussed in earlier chapters: (1) business growth and expansion, (2) strategic cost management, (3) operations, logistics, and technology management, (4) employee development and satisfaction, and (5) resource management.

Tables 4.7 through 4.26 cover several parameters to be monitored and measured as a company proceeds toward a strategy execution that will allow it to meet its vision. All the tables can be modified to perform the required analysis based on products rather than factories or operating divisions.

Table 4.2 Business Growth and Expansion Monitoring

No.	Performance Statement	Periodicity
1	Routine Business Development	Quarterly
2	New Business Development	Every 6 months
3	Consolidated Statement of Business Development	Yearly

Table 4.3 Strategic Cost Management Monitoring

No.	Performance Statement	Periodicity
1	Operations Cost Statement	Quarterly
2	Cost Reduction and Cost Competitiveness	Every 6 months
3	Strategic Cost Management	Every 6 months
4	Application of Activity-Based Costing	Quarterly

Table 4.4 Operations, Logistics, and Technology Management Monitoring

No.	Performance Statement	Periodicity
1	Quantitative Operations Statement	Quarterly
2	Strategic Aspects of Operations Management	Every 6 months
3	Consolidated Statement of Operations Management	Yearly

Table 4.5 Employee Development and Satisfaction Monitoring

No.	Performance Statement	Periodicity
1	Knowledge Acquisition	Quarterly
2	Knowledge Application	Every 6 months
3	Strategic Issues in Knowledge Management	Every 6 months
4	Routine Productivity Follow-up	Quarterly
5	Productivity Enhancement Efforts and Results	Every 6 months
6	Employee Satisfaction	Every 6 months

Table 4.6 Resource Management Monitoring

No.	Performance Statement	Periodicity
1	Asset Utilization Efficiency	Quarterly
2	Routine Capital Expenditure	Quarterly
3	Special Capital Expenditure	Every 6 months
4	Strategic Issues in Fund Mobilization and Deployment	Every 6 months

Business Growth and Expansion Monitoring Tools

Table 4.7 Statement 1: Routine Business Development (Quarterly)

No.	Monitoring Parameter	Optimistic Expectation (a)	Moderate Expectation (b)	Actual Performance (c)	Deviation 1 (a) – (c)	Deviation 2 (b) – (c)	Reasons
1	Sales growth						
2	Change in market share						
3	Creation of new market						
4	Promotional costs/sales						
5	Incremental promotional cost/ incremental sales						
6	Incremental sales per employee						
7	Incremental gross profit/ incremental sales						

Table 4.8 Statement 2: New Business Development (Six-Month Intervals)

No.	Monitoring Parameter	Optimistic Expectation (a)	Moderate Expectation (b)	Actual Performance (c)	Deviation 1 (a) – (c)	Deviation 2 (b) – (c)	Reasons
1	Commercialization of new product, service, or operation						
2	Phasing completion of new project						
3	Phasing capital cost of new project						
4	Operating return on investment (ROI) of new business						
5	Investment turnover of new business						
6	Operating profitability of new business						
7	Complementary impact of new business on present business						
8	Marginal or incremental employee efficiency of new business						
9	Incremental or marginal cost of capital for new business						
10	Management or time-based ROI of new business						

Table 4.9 Statement 3: Consolidated Statement of Business Development (Yearly)

No.	Monitoring Parameter	Optimistic Expectation (a)	Moderate Expectation (b)	Actual Performance (c)	Deviation 1 (a) – (c)	Deviation 2 (b) – (c)	Reasons
1	Consolidated ROI of new business						
2	Incremental employee efficiency of new business						
3	Consolidated market share						
4	Consolidated investment turnover of total business						
5	Synergic advantage of new business						
6	Marginal or incremental profitability of new business						
7	Impact of capital cost of new business on cash flow management						
8	Time-based ROI of total business						

Strategic Cost Management Monitoring Tools

Table 4.10 Statement 1: Operations Cost Management (Quarterly)

No.	Monitoring Parameter	Optimistic Expectation (a)	Moderate Expectation (b)	Actual Performance (c)	Deviation 1 (a) – (c)	Deviation 2 (b) – (c)	Reasons
1	Operating costs/sales						
2	Value added/cost of value addition						
3	Incremental cost of new, diversified, or altered operation						
4	Employee cost/operations cost						
5	Promotional cost/operational cost						
6	Inventory carrying cost/operations cost						
7	Cost of credit/total sales						
8	Incremental gross profit per dollar of incremental operating cost						
9	Net cost advantage resulting from improved operational synergy						
10	Cost of managing government issues						

Table 4.11 Statement 2: Cost Reduction and Cost Competitiveness (Six-Month Intervals)

No.	Monitoring Parameter	Optimistic Expectation (a)	Moderate Expectation (b)	Actual Performance (c)	Deviation 1 (a) – (c)	Deviation 2 (b) – (c)	Reasons
1	New cost competitiveness through segmental improvement						
2	Reduction in routine costs						
3	Reduction in development cost without diluting quality or scope of development						
4	Employees' contributions to cost competitiveness						
5	Systems' contributions to cost competitiveness						
6	External factors' contributions to cost competitiveness						
7	Impact of cost competitiveness on market leadership or acquisition of new market segment						
8	Cost reduction through economy of scale						
9	Impact of entrepreneurial pay scheme on cost competitiveness						

Table 4.12 Statement 3: Strategic Cost Management (Six-Month Intervals)

No.	Monitoring Parameter	Optimistic Expectation (a)	Moderate Expectation (b)	Actual Performance (c)	Deviation 1 (a) – (c)	Deviation 2 (b) – (c)	Reasons
1	Impact of organization structure changes on cost efficiencies (e.g., impact of more empowerment)						
2	Cost in product/cost on product						
3	Committed cost/total cost						
4	Variable wage cost/total wage cost						
5	Reduction in non-value added costs						
6	Reduction in opportunity cost						
7	Cost advantage through acquisition of strategically located facilities						
8	Cost milestones for new product at various phases of its life cycle: Launch Take-off Consolidation Leadership						
9	Reduction in sunk cost						
10	Reduction in weighted cost of capital through financial restructuring						

Table 4.13 Statement 4: Application of Activity-Based Costing (Quarterly)

No.	Monitoring Parameter	Optimistic Expectation (a)	Moderate Expectation (b)	Actual Performance (c)	Deviation 1 (a) – (c)	Deviation 2 (b) – (c)	Reasons
1	Designation of assets for independent functions or activities						
2	Conversion of common cost into identifiable costs						
3	Alteration in value addition resulting from activity						
4	Cost of alteration in value addition						
5	Product costing accuracy using ABC						
6	System improvisation to suit ABC						

Operations, Logistics, and Technology Management Monitoring Tools

Table 4.14   Statement 1: Quantitative Operations Statement (Quarterly)

No.	Monitoring Parameter	Optimistic Expectation (a)	Moderate Expectation (b)	Actual Performance (c)	Deviation 1 (a) – (c)	Deviation 2 (b) – (c)	Reasons
1	Operating productivity of business unit						
2	Operating productivity of each functional group						
3	Operating costs/sales						
4	Improvement in intragroup operational synergy						
5	Operating ROI of business unit						
6	Operating ROI of each profit center						
7	Incremental cost for achieving incremental operating productivity						
8	Core operating cost/support operating costs						
9	Incremental gross profit per unit of quality improvement						
10	Impact of operational corrections on costs						

Note: Operating productivity of each functional group and operating ROI of each profit center may be shown separately.

Table 4.15   Statement 2: Strategic Aspects of Operations Management (Six-Month Intervals)

No.	Monitoring Parameter	Optimistic Expectation (a)	Moderate Expectation (b)	Actual Performance (c)	Deviation 1 (a) – (c)	Deviation 2 (b) – (c)	Reasons
1	Strategic acquisitions of new capacities or processes						
2	Enhancement of competitive advantages from process technologies						
3	Improvement in government relations						
4	Improvement in operational logistics arising from improvement of infrastructure						
5	Conversion of departments, plants, core functions, or support functions into profit centers via application of concepts like activity-based costing, economic value added (EVA), etc.						
6	Impact of new entrepreneurial pay design on operational efficiency						
7	Sustainable financial viability of new and old operations						
8	Improvement in self-supportive operations of subsidiaries						
9	Improvement in earnings and operational sustainability of diversified businesses						

Table 4.16 Statement 3: Consolidated Statement of Operations Management (Yearly)

No.	Monitoring Parameter	Optimistic Expectation (a)	Moderate Expectation (b)	Actual Performance (c)	Deviation 1 (a) – (c)	Deviation 2 (b) – (c)	Reasons
1	Reduction in operating costs						
2	Improvement in operating productivity of business unit or functional group						
3	Improvement in operational logistics and synergy						
4	Strategic advantage of new, diversified, or altered operations						
5	Improvement in government relations						
6	Consolidation of core operations with subsidiary operations						

Employee Development and Satisfaction Monitoring Tools

Table 4.17 Statement 1: Knowledge Acquisition (Quarterly)

No.	Monitoring Parameter	Optimistic Expectation (a)	Moderate Expectation (b)	Actual Performance (c)	Deviation 1 (a) – (c)	Deviation 2 (b) – (c)	Reasons
1	Identification of new competencies based on business development plan						
2	Identification of competence gaps						
3	Knowledge sourcing for: New skills New systems New business segments						
4	Cost of knowledge acquisition: One-time costs Recurring costs Unproductive costs						
5	Employee contributions to knowledge acquisition						

Table 4.18 Statement 2: Knowledge Application (Six-Month Intervals)

No.	Monitoring Parameter	Optimistic Expectation (a)	Moderate Expectation (b)	Actual Performance (c)	Deviation 1 (a) – (c)	Deviation 2 (b) – (c)	Reasons
1	Completion of gap bridging						
2	Cost of gap bridging						
3	Gross incremental revenue or benefit from new competence application						
4	Sustainable change in productivity through knowledge application						

Table 4.19 Statement 3: Strategic Issues in Knowledge Management (Six-Month Intervals)

No.	Monitoring Parameter	Optimistic Expectation (a)	Moderate Expectation (b)	Actual Performance (c)	Deviation 1 (a) – (c)	Deviation 2 (b) – (c)	Reasons
1	Intragroup and interdepartment knowledge sharing efforts						
2	Knowledge sharing facilitation for employees						
3	Incremental advantage/incremental cost of knowledge management						
4	Absorption of cost of knowledge management: Cost capitalization Annual absorption or amortization						
5	Sourcing of knowledge for potential use Quantum Cost						
6	Sale of knowledge for additional revenue						
7	Market leadership through knowledge management						
8	Employee retention through knowledge management						
9	Impact of entrepreneurial pay package on knowledge management						

Table 4.20 Statement 4: Routine Productivity Follow-Up (Quarterly)

No.	Monitoring Parameter	Optimistic Expectation (a)	Moderate Expectation (b)	Actual Performance (c)	Deviation 1 (a) – (c)	Deviation 2 (b) – (c)	Reasons
1	Average employee productivity						
2	Average productivity of new employees						
3	Operational or plant productivity						
4	Reduction in normal productivity losses						
5	Reduction in abnormal productivity losses						
6	Reduction in cost per unit of routine productivity						
7	Improvement in routine productivity through logistics or technology improvements						
8	Employee contributions to routine productivity improvement						
9	Impact of change of business model on routine productivity						
10	Improvement in routine productivity through improved management of external variables (e.g., government and supplier relations)						

Table 4.21 Statement 5: Productivity Enhancement Efforts and Results (Six-Month Intervals)

No.	Monitoring Parameter	Optimistic Expectation (a)	Moderate Expectation (b)	Actual Performance (c)	Deviation 1 (a) – (c)	Deviation 2 (b) – (c)	Reasons
1	Productivity improvement through better technology						
2	Productivity improvement through additional employee efforts						
3	Cost of improving productivity: Capital costs Recurring costs						
4	Productivity improvement through structural changes						
5	Gross incremental revenue from incremental productivity						
6	Improvement in productivity measurement						
7	Impact of entrepreneurial pay package on productivity improvement						

Table 4.22 Statement 6: Employee Satisfaction (Six-Month Intervals)

No.	Monitoring Parameter	Optimistic Expectation (a)	Moderate Expectation (b)	Actual Performance (c)	Deviation 1 (a) – (c)	Deviation 2 (b) – (c)	Reasons
1	Employees left/total employees						
2	New employees/total employees						
3	Employee satisfaction (scale of 0 to 5*) with: Pay structure; Incentives; Promotions; Annual pay revision; Training; Target setting; Performance reviews; Working conditions; Empowerment; Guidance and support from superiors; Other amenities						

*0–5 scale: 0 = Nil, 1 = Very poor, 2 = Poor, 3 = Average, 4 = Good, 5 = Very good.

Resource Management Monitoring Tools

Table 4.23 Statement 1: Asset Utilization Efficiency (Quarterly)

No.	Monitoring Parameter	Optimistic Expectation (a)	Moderate Expectation (b)	Actual Performance (c)	Deviation 1 (a) – (c)	Deviation 2 (b) – (c)	Reasons
1	Fixed-asset turnover						
2	Working capital turnover						
3	Operating or plant capacity usage						
4	Employee capacity usage						
5	Marketing capacity usage						
6	Cost of incremental asset utilization						
7	Impact of logistic improvement on asset utilization						
8	Impact employee productivity improvement on asset utilization						
9	Incremental fixed-asset turnover						
10	Incremental working capital turnover						

Table 4.24 Statement 2: Routine Capital Expenditure (RCE) (Quarterly)

No.	Monitoring Parameter	Optimistic Expectation (a)	Moderate Expectation (b)	Actual Performance (c)	Deviation 1 (a) – (c)	Deviation 2 (b) – (c)	Reasons
1	ROI or internal rate of return (IRR) on routine capital expenditures						
2	Cash outflow efficiency (liquidity management) of RCE						
3	Reduction of RCE through better asset use or operations management						

Table 4.25   Statement 3: Special Capital Expenditures (SCE) (Six-Month Intervals)

No.	Monitoring Parameter	Optimistic Expectation (a)	Moderate Expectation (b)	Actual Performance (c)	Deviation 1 (a) − (c)	Deviation 2 (b) − (c)	Reasons
1	ROI or IRR on special capital expenditures						
2	Efficient phase-wise completion of SCE						
3	Impact of SCE on combined ROI of business unit						
4	Liquidity management of SCE						
5	Impact of SCE on employee productivity						
6	Absorption of SCE: Amortization Depreciation						
7	Efficiency in use of assets or facilities purchased through SCE						
8	Opportunity cost of funds invested in SCE						
9	Complementary impact of SCE on other projects, functions, or businesses						
10	Time-based ROI on SCE						

Table 4.26 Statement 4: Strategic Issues in Fund Mobilization and Deployment (Six-Month Intervals)

No.	Monitoring Parameter	Optimistic Expectation (a)	Moderate Expectation (b)	Actual Performance (c)	Deviation 1 (a) – (c)	Deviation 2 (b) – (c)	Reasons
1	Incremental weighted cost of incremental funds raised						
2	Incremental ROI on incremental funds deployed						
3	Phase-wise efficiency of fund raising and deployment						
4	Efficiency in managing following risks during fund-raising and fund deployment Administration Liquidity Interest rates Foreign exchange Commercial activities Legal issues Portfolio						
5	Flexibility in fund management: Conversion Reduction Endorsement						
6	Efficient use of funds by each business unit, function, or department operating as a profit center						

Chapter 5

Measurement of Strategy Execution and Financial Monitoring

Pyramid Movement of ROI for Strategy Execution

The ultimate objective of strategic planning is wealth maximization and it requires every strategy to offer optimal profit directly or indirectly. Therefore, the execution process of each strategy must be monitored financially and its ultimate impact should be measured financially. Strategy execution is achieved by completing a set of strategy-related functions or activities. Hence, a mechanism is required to view each function or activity of an organization as a business unit or profit center—notionally or in practice.

For example, a company may treat its research and development unit as an independent business that is supposed to earn profits directly or indirectly or it may consider its repair and maintenance (R&M) function as a profit center. The concept of profit center is also known as "responsibility accounting."

A company's R&M division is normally considered a cost center because it incurs costs to provide repairs and maintenance services across operations. If the division is allowed to charge a notional price inclusive of profit to its internal customers, it becomes a profit center. If the divisional head is empowered to determine capital investments of the division and earn a notional return on invesment (ROI), the division becomes an investment center. Figure 5.1 illustrates the ascending degrees of empowerment of a functional division.

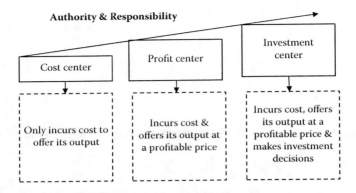

Figure 5.1 Empowerment of functional division.

Financial monitoring and measurement techniques require benchmarking of every cost and profit item. This compels an organization to watch market conditions closely and expect its employees to perform like entrepreneurs. As noted earlier, a strategic plan must focus on ROI. Financial monitoring and measurement require ROI to trickle down to the levels of divisions and subdivisions. This is known as pyramid movement, based on the descending order of strategy execution as depicted in Figure 5.2.

Financial Impacts of Strategy Execution

Every strategy exerts an impact on an organization visibly or invisibly over the short run and/or the long run. These impacts must be monitored and measured and ultimately utilized in various reports. Every strategy impacts financial records such as a company's income statements, cash flow statements, and balance sheets. A strategic plan should be converted into a series of annual financial budgets by projecting the financial impacts of the strategy execution for the coming year, as illustrated in Figure 5.3. Let us look at two examples that analyze the financial impact of a strategy execution.

Example 5.1

Strategy—Purchase a patented design of a critical production process for a one-time (capital) cost of $36 million. The purchase should bring down the present material consumption per unit of product by 5%. The present

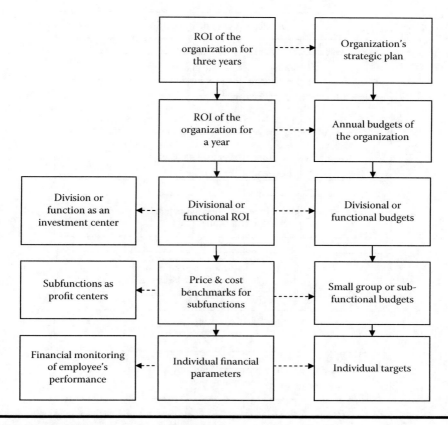

Figure 5.2 Pyramid movement of ROI.

raw material cost incurred per unit of product is $100. The patented design will have an economic life of 6 years. The capital cost is to be funded by 10% borrowing. The loan is to be repaid in six equal annual installments. Present annual sales volume is 5 million units.

Impact Areas—
  Cost management (reduced material cost)
  Operations and logistics (patented design)
  Resource management (funding through loan)

Impacts of Strategy—The impacts are shown in Tables 5.1 through 5.3. Relevant notes about the statements follow Table 5.3.

Example 5.2

Strategies—One hundred workers will receive one-time compensation of $10,000 each. This compensation should be written off in 5 years. Each one

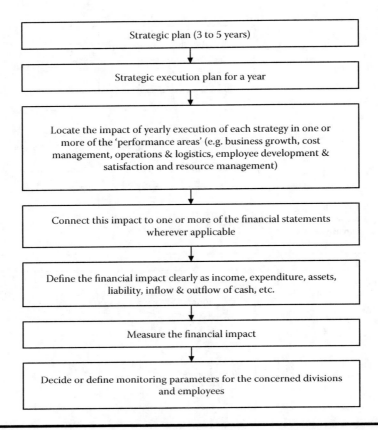

Figure 5.3 Conversion of strategic plan into annual budget.

of the fifty managers will receive 100 company shares at no cost. The face value of a share is $2. Two hundred supervisors will be offered 1-week training programs. The cost to the company will be $140 per person; cost should be written off in the year incurred.

Impact Areas—
    Cost management (compensation and training cost)
    Employee development and satisfaction (impacts of share issue and training on productivity and satisfaction)
    Resource management (increase in share capital based on issue of free shares to employees)

Impacts of Strategy—The impacts are shown in Tables 5.4 through 5.6. Relevant notes about the statements follow Table 5.6.

Table 5.1 Impact on Income Statement for Year

Description	$ (million)	$ (million)
Sales (5 million units)		xx
Less Operating Cost		
Material	500*	
Less Savings ↓	–25*	
Remuneration	x	
Factory Overheads	x	
Administrative Overheads	x	
Sales Overheads	x	
Depreciation	x	
Add Amortization of Patented Design ↑	+6*	xx + (19) ↓
Profit before Income Taxes (PBIT)		xx
Less Interest	x	
Add 10% on Loan ↑	3.6*	xx + (3.6) ↑
Profit before Taxes (PBT)		xx (15.4) ↑
Less Taxes		xx
Profit after Taxes (PAT)		xx
Less Dividend		xx
Retained Profit		xx ↑

*Items impacted by execution of strategy.

Financial Monitoring

In earlier chapters, we defined five performance areas for strategic planning:

- Business growth and expansion
- Strategic cost management
- Operations, logistics, and technology management
- Employee development and satisfaction
- Resource management

The areas are interdependent and hence their impact is collective. Therefore, the financial impact of growth in each performance area

Table 5.2 Impact on Cash Flow Statement for Year

Description	$ (million)
Cash Inflows	
Opening Balance	xx
Sales	xx
Receipt of Loan Proceeds	36*
Total Cash Inflows (a)	xxx
Cash Outflows	
Operating Costs	
Material	475*
Remuneration	xx
Factory Overheads	xx
Administrative Overheads	xx
Sales Overheads	xx
Interest on Old Loan	xx
Interest on New Loan	3.6*
Tax	xx*
Dividend	xx
Payment of Loan	6*
Purchase of Patented Design	36*
Total Cash Outflows (b)	xxx
Closing Balance (a) – (b)	xx

*Items impacted by execution of strategy.

must be monitored through collective parameters. Monitoring should not (and to some extent cannot) be a separate exercise for a single growth area.

To monitor financial impacts of a strategy implementation, relevant financial data can be sourced from an organization's present accounting system. Monitoring of various parameters should be performed at various time periods based on need. For example, monitoring the costs of overtime pay to workers may be monitored weekly, whereas performance-based variable payments to executives may be monitored monthly or quarterly. CEO's performance was monitored once or

Table 5.3 Impact on Balance Sheet

Fund Source		$ (million)	Investment of Funds		$ (million)
Share Capital		xx	Fixed Assets after		
Reserves		xx	Depreciation	xx	
Borrowed Funds	xx		Add Patented Design	36	
Add New Loan	36*		Less Amortization	–6*	xx + 30
Less Repayment	–6*	xx + 30			
			Working Capital (a) – (b)		xx
			Current Assets (a)		
			Cash* ↑		
			Stock of Raw Material* ↓		
			Receivables		
			Stock of Finished Goods* ↓		
			Less Current Liabilities (b)		
			Creditors* ↓		
			Bank Overdraft* ↓		
Total		xxx	Total		xxx

Notes on Income Statement, Cash Flow Statement, and Balance Sheet Computations:
↑ indicates increase; ↓ indicates decrease.

Original material cost (without saving in consumption) = 5 million units @ $100 = $500 million, creating a 5% saving ($25 million) in material consumption. Cost of patented design = $36 million. Economic life of patent = 6 years. Yearly amortization = $6 million ($36 million/6 years). Amortization is an operating cost.

Patent purchase funded by loan at 10% interest per annum. Interest on loan = 10% on $36 million = $3.6 million per annum.

The benefit of saving in material cost ($25 million) exceeds the total cost of amortization ($6 million) and interest on loan ($3.6 million), resulting in net increase in profit.

Saving on cash outflow through material purchases = $25 million. This exceeds the cash outflow for loan repayment ($6 million) and interest payment ($3.6 million). As a net result, closing cash balance at year end should increase.

Because of saving in material consumption, the values of closing stocks of raw material and finished goods should decline, thus reducing working capital.

Savings in material purchase costs should reduce average creditors item (amounts payable to suppliers).

Overall increase in cash flow should reduce bank overdraft item.

*Items impacted by execution of strategy.

152 ■ *Strategic Planning, Execution, and Measurement*

Table 5.4 Impact on Income Statement for Year

Description	$ (million)	$ (million)
Sales (5 million units)		xx
Less Operating Costs		
Remuneration	xx	
Add Written-off Workers' Compensation	0.2*	
Add Free Shares Issued to Managers	0.01*	
Add Supervisor Training Costs	0.028*	
Other Operating Costs	xx	xx + 1.19 ↑
Profit before interest & tax (PBIT)		xx ↓
Less Interest		xx
Profit before tax (PBT)		xx ↓
Less Tax		xx ↓
Profit after tax (PAT)		xx ↓
Less Dividend		xx
Retained Profit		xx ↓

*Items impacted by execution of strategy.

twice annually but current practice is to monitor it quarterly. If business continues to become more competitive, CEO's performance may even be monitored monthly.

Employees at different levels report on different parameters. For example, systemic employees may be accountable for value addition costs or production-based costs of incentives and should report on just a few parameters. The number of reporting parameters increases for knowledge employees such as team or group leaders (supervisors and officers) and the number increases further for decision-making employees such as divisional, functional, or location heads. Therefore, the financial parameters presented in this chapter mainly cover the four levels of employees based on the levels of their responsibilities as shown in Figure 5.4.

Common Characteristics of Financial Parameter Tables

Tables 5.7 through 5.20 detail financial and productivity parameters for all classes of employees from entry-level workers through executives.

Table 5.5 Impact on Cash Flow Statement for Year

Description	$ (million)
Cash Inflows	
Opening Balance	xx
Sales	xx
Total Cash Inflows (a)	xxx
Cash Outflows	
Operating Costs	
Remuneration	xx
Workers' Compensation	1.000*
Supervisor Training Costs	0.028*
Other Operating Costs, Interest, and Taxes	xx
Dividend	xx
Total Cash Outflows (b)	xx
Closing Balance (a) – (b)	xx ↓

*Items impacted by execution of strategy.

Table 5.6 Impact on Balance Sheet

Fund Source		$ (million)	Investment of Funds	$ (million)
Share Capital	xx		Fixed Assets	xx
Add Free Shares Issued to Managers	(0.01)*	xx + 0.01	Working Capital (a) – (b)	
			Current Assets (a):	xx
Reserves		xx	Cash* ↓	
Borrowed Funds		xx	Other Current Assets	
			Less Current Liabilities (b):	
			Workers' Compensation (not written off)	0.8*
Total		xxx	Total	xxx

Notes on Income Statement, Cash Flow Statement, and Balance Sheet Computations:
  ↑ indicates increase; ↓ indicates decrease.
  Total workers' compensation = $1 million (100 workers × $10,000) to be written off over 5 years = $200,000 annually. Remaining $800,000 not written off is shown as an asset on the balance sheet.

*Items impacted by execution of strategy.

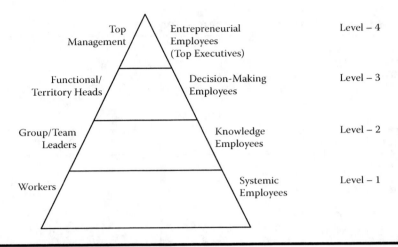

Figure 5.4 Levels of employees.

Productivity is defined as a function of (1) cost, (2) volume, (3) time, (4) quality, (5) safety, and (6) acceptability. All the parameters are categorized into four parts and are abbreviated as shown below:

Cost management inquiry (CMI)
Employee productivity inquiry (EPI)
Operations management inquiry (OMI)
Resource management inquiry (RMI)

Suggested weights for the parameters are scaled as follows:

Very high–5
High–4
Considerable–3
Low–2
Negligible–1

Level 1: Systemic Employees

Table 5.7 Production (Operating) Employees

No.	Parameter	Periodicity of Measurement	Suggested Weight and Category
1	Employee's total cost per unit of activity, process, or operation (total cost = identifiable costs and common costs apportioned to employee activity)	Weekly	4 CMI
2	Employee's identifiable or direct cost per unit of employee activity	Weekly	3 CMI
3	Variable (performance-based) wage/total wage	Monthly	4 EPI
4	Overtime wages/total wage	Monthly	2 EPI
5	Idle time wage/total wage	Monthly	2 EPI
6	Dearness allowance/total wage	Monthly	1 CMI
7	Waste cost per unit of activity/identifiable cost per unit of employee's activity	Weekly	2 EPI
8	Controllable cost per unit of employee's activity, process, or operation	Weekly	3 CMI
9	Cost of value addition per unit of employee's activity, process, or operation	Weekly	4 CMI
10	Total cost per unit of employee's activity, operation, or process/total cost per unit of complete product	Weekly	2 OMI
11	Employee's value addition cost per unit of activity/total value addition cost per unit of complete product	Weekly	3 OMI
12	Fixed cost per unit of employee's activity	Monthly	2 CMI
13	Total cost per unit of employee's activity/market price per unit of his activity	Monthly	3 CMI
14	Identified cost per unit of employee's activity/market price per unit of activity	Monthly	2 CMI

(Continued)

Table 5.7 (*Continued*)   Production (Operating) Employees

No.	Parameter	Periodicity of Measurement	Suggested Weight and Category
15	Employee's identifiable cost per unit of group's activity/total identifiable cost per unit of group's activity	Monthly	2 CMI
16	Total cost per unit of group's activity/total cost per unit of complete product	Monthly	3 CMI
17	Total identifiable cost per unit of group's activity/total identifiable cost per unit of complete product	Monthly	2 CMI
18	Total cost per unit of group's activity/market price per unit of group's activity	Monthly	2 CMI
19	Material cost per unit of employee's activity/total cost per unit of employee's activity	Weekly	2 CMI
20	Employee's allowances and perquisites/total salary	Monthly	2 CMI
21	Retirement benefits contributed by employer/total salary	Monthly	2 CMI
22	Cash salary as percentage of total salary	Monthly	2 EPI
23	Taxable salary as percentage of total salary	Monthly	2 EPI
24	Percentage salary increase (excluding dearness allowance)/percentage increase in employee productivity	Annually/six monthly	4 EPI
25	Employee's salary/average salary in region	Six monthly	2 EPI
26	Employee's salary/average salary in industry	Six monthly	2 EPI
27	Employee's salary/average salary in organization	Six monthly	3 EPI
28	Employee's salary/salary of boss	Six monthly	2 EPI
29	Employee's salary/highest salary in organization	Six monthly	2 EPI

Table 5.7 (*Continued*)   Production (Operating) Employees

No.	Parameter	Periodicity of Measurement	Suggested Weight and Category
30	Actual salary per degree or point or quotient of quality/targeted salary per degree of quality	Quarterly	3 EPI
31	Employee's total cost to company/total hiring cost of contract or temporary worker	Six monthly	3 OMI
32	Employee's ownership rights or shares/average ownership shares per employee of organization	Annually	2 EPI
33	Employee's ownership shares/highest number of shares owned by employee of organization	Annually	3 EPI
34	Employee's ownership shares/average ownership shares per employee in industry or rival company	Annually	3 EPI
35	Average rate of increase in productivity during past 3 or 5 years/average rate of increase in total salary during last 3 or 5 years	Annually	3 EPI
36	Average rate of increase in total salary during past 3 or 5 years/average rate of increase in total average salary in organization or same class of worker	Annually	3 EPI
37	Average rate of increase in employee productivity over past 3 or 5 years/average rate of increase in average productivity of organization or same class of worker	Annually	4 EPI
38	Salary of idle time including holidays as a percentage of total salary/average salary of average idle time as percentage of average total salary	Quarterly	3 EPI
39	Salary per degree of employee's productivity/average salary per degree of productivity of newer workers in same category	Quarterly	3 EPI

(*Continued*)

Table 5.7 (*Continued*)   Production (Operating) Employees

No.	Parameter	Periodicity of Measurement	Suggested Weight and Category
40	Average rate of salary increase in past 6 years/average rate of increase in operating profit of organization during past 6 years	Annually	4 EPI
41	Cost of employee training and development in areas of skills, attitude, and knowledge	Annually	2 EPI
42	Cost of employee training and development per unit of extra productivity/average cost of training and development per worker per unit of average extra productivity	Annually	3 EPI
43	Cost of employee's annual training and development/value addition achieved by employee in past year	Annually	3 EPI
44	Cost of welfare for employee/employee's total salary	Annually	2 CMI
45	Take-home salary/total salary	Quarterly	2 EPI
46	Average rate of increase in employee's productivity during past 6 years/average rate of increase in cost of infrastructure in past 6 years	Annually	3 EPI
47	Cost per extra unit made by employee/average cost per unit incurred by employee	Annually	3 CMI
48	Average rate of increase in dearness allowance in past 6 years/average rate of inflation during past 6 years	Annually	2 CMI
49	Average rate of increase in employee's savings over past 6 years/average rate of increase in salary during past 6 years	Annually	3 EPI
50	Cost of abnormal waste caused by employee/cost of normal waste incurred by employee	Monthly	2 CMI

Table 5.8 Marketing Workers (Sales People, Distribution and Warehouse Workers, and Sales Office Support Staff)

No.	Parameter	Periodicity of Measurement	Suggested Weight and Category
1	Total cost per unit of employee's activity, process, or operation (total cost = identifiable costs and common costs apportioned to employee's activity)	Weekly	4 CMI
2	Identifiable or direct cost per unit of employee activity	Weekly	3 CMI
3	Employee's variable (performance-based) wage/total wage	Monthly	4 EPI
4	Overtime wage/total wage	Monthly	2 EPI
5	Idle time wage/total wage	Monthly	2 EPI
6	Dearness allowance/total wage	Monthly	1 CMI
7	Cost of waste per unit of employee's activity/identifiable cost per unit of activity	Weekly	2 EPI
8	Controllable cost per unit of employee's activity, process of operation	Weekly	3 CMI
9	Cost of value addition per unit of employee's activity, process, or operation	Weekly	4 CMI
10	Total cost per unit of employee's activity, operation or process/total cost per unit of complete product	Weekly	2 OMI
11	Value addition cost per unit of employee's activity/total value addition cost per unit of complete product	Weekly	3 OMI
12	Fixed cost per unit of employee's activity	Monthly	2 CMI
13	Total cost of employee's activity/market price per unit of activity	Monthly	3 CMI
14	Identified cost per unit of employee's activity/market price per unit of activity	Monthly	2 CMI

(Continued)

Table 5.8 (*Continued*)   Marketing Workers (Sales People, Distribution and Warehouse Workers, and Sales Office Support Staff)

No.	Parameter	Periodicity of Measurement	Suggested Weight and Category
15	Employee's identifiable cost per unit of group's activity/total identifiable cost per unit of group's activity	Monthly	2 CMI
16	Total cost per unit of group's activity/total cost per unit of complete product	Monthly	3 CMI
17	Total identifiable cost per unit of group's activity/total identifiable cost per unit of complete product	Monthly	2 CMI
18	Total cost per unit of group's activity/market price per unit of group's activity	Monthly	2 CMI
19	Employee's allowances and perquisites/total salary	Monthly	2 CMI
20	Retirement benefits contributed by employer/total salary	Monthly	2 CMI
21	Cash salary as percentage of total salary	Monthly	2 EPI
22	Taxable salary as percentage of total salary	Monthly	2 EPI
23	Percentage salary increase (excluding dearness allowance)/percentage increase in productivity	Annually/six monthly	4 EPI
24	Employee's salary/average salary in region	Six monthly	2 EPI
25	Employee's salary/average salary in industry	Six monthly	2 EPI
26	Employee's salary/average salary in organization	Six monthly	3 EPI
27	Employee's salary/salary of boss	Six monthly	2 EPI
28	Employee's salary/highest salary in organization	Six monthly	2 EPI
29	Total cost to company/total cost of hiring contract or temporary worker	Six monthly	3 OMI

Table 5.8 (*Continued*) Marketing Workers (Sales People, Distribution and Warehouse Workers, and Sales Office Support Staff)

No.	Parameter	Periodicity of Measurement	Suggested Weight and Category
30	Employee's ownership rights or shares/average ownership shares per employee of organization	Annually	2 EPI
31	Employee's ownership shares/highest number of shares owned by employee of organization	Annually	3 EPI
32	Employee's ownership shares/average ownership shares per employee in industry or rival company	Annually	3 EPI
33	Average rate of increase in employee productivity during past 3 or 5 years/average rate of increase in total salary during past 3 or 5 years	Annually	3 EPI
34	Average rate of increase in total salary during past 3 or 5 years/average rate of increase in total average salary in organization same class of worker	Annually	3 EPI
35	Average rate of increase in employee productivity in past 3 or 5 years/average rate of increase in productivity of organization or same class of worker	Annually	4 EPI
36	Employee's idle time salary including holidays as percentage of total salary/average idle time salary as percentage of average total salary	Quarterly	3 EPI
37	Employee's salary per degree of productivity/average salary per degree of productivity of newer workers in category	Quarterly	3 EPI
38	Average rate of salary increases in past 6 years/average rate of increase in operating profit of organization over past 6 years	Annually	4 EPI

(*Continued*)

Table 5.8 (*Continued*) Marketing Workers (Sales People, Distribution and Warehouse Workers, and Sales Office Support Staff)

No.	Parameter	Periodicity of Measurement	Suggested Weight and Category
39	Cost of employee's training and development in areas of skills, attitude, and knowledge	Annually	2 EPI
40	Cost of employee's training and development per unit of his or her extra productivity/average cost of training and development per worker per unit of average extra productivity	Annually	3 EPI
41	Cost of employee's annual training and development/value additions produced by employee in past year	Annually	3 EPI
42	Cost of welfare for employee/total salary	Annually	2 CMI
43	Take-home salary/total salary	Quarterly	2 EPI
44	Average rate of increase in employee's productivity over past 6 years/average rate of increase in cost of infrastructure during past 6 years	Annually	3 EPI
45	Cost per extra unit produced by employee/average cost per unit incurred by employee	Annually	3 CMI
46	Average rate of increase in employee's dearness allowance during past 6 years/average rate of inflation during past 6 years	Annually	2 CMI
47	Average rate of increase in employees savings during past 6 years/average rate of salary increases in past 6 years	Annually	3 EPI
48	Sales volume achieved per customer	Monthly	3 OMI
49	Sales value achieved per customer	Monthly	3 OMI
50	Employee's sales overhead as percentage of sales value he or she achieved	Monthly	3 CMI
51	Average sales per working day	Weekly	3 EPI

Table 5.8 (*Continued*) Marketing Workers (Sales People, Distribution and Warehouse Workers, and Sales Office Support Staff)

No.	Parameter	Periodicity of Measurement	Suggested Weight and Category
52	Employee's bad debts/employee's sales	Monthly	3 CMI
53	Employee's debt turnover ratio	Quarterly	3 OMI
54	Employee's cost of credit/employee's sales	Monthly	3 CMI
55	Sales target achievement	Weekly	5 EPI
56	Employee's sales/employee's cost to company	Quarterly	3 EPI
57	Percentage growth in employee market share/percentage growth in market	Quarterly	3 OMI
58	Percentage usage of the warehousing capacity	Monthly	2 OMI
59	Percentage use of transportation facility	Monthly	3 OMI
60	Sales overhead versus production overhead	Quarterly	2 CMI
61	Employee's market share versus competitor's market share	Quarterly	3 OMI
62	Employee's group's sales administration costs/group's total sales overhead	Quarterly	2 CMI
63	Trend of change in employee's market share/trend of change in market during last 3 to 5 years	Annually	4 OMI
64	Employee's profit achieved/employee's profit targeted	Quarterly	4 EPI
65	Employee's cash profit/employee's total profit	Quarterly	3 OMI
66	Identifiable profit achieved/identifiable profit targeted	Quarterly	3 EPI
67	Employee's inventory turnover ratio	Quarterly	3 EPI

Table 5.9 Purchasing (Supply Chain) Workers (Clerks, Store Keepers, Helpers, Drivers, Security Guards)

No.	Parameter	Periodicity of Measurement	Suggested Weight and Category
1	Total cost per unit of employee's activity, process, or operation (total cost = identifiable costs and common costs apportioned to employee's activity)	Weekly	4 CMI
2	Identifiable or direct cost per unit of employee's activity	Weekly	3 CMI
3	Variable (performance-based wage/total wage)	Monthly	4 EPI
4	Overtime wage/total wage	Monthly	2 EPI
5	Idle time wage/total wage	Monthly	2 EPI
6	Dearness allowance/total wage	Monthly	1 CMI
7	Cost of waste per unit of employee's activity/identifiable cost per unit of activity	Weekly	2 EPI
8	Controllable cost per unit of employee's activity, process, or operation	Weekly	3 CMI
9	Cost of employee's value addition per unit of activity, process, or operation	Weekly	4 CMI
10	Total cost per unit of employee's activity, operation, or process/total cost per unit of complete product	Weekly	2 OMI
11	Value addition cost per unit of employee's activity/total value addition cost per unit of complete product	Weekly	3 OMI
12	Fixed cost per unit of employee's activity	Monthly	2 CMI
13	Total cost per unit of employee's activity/market price per unit of employee's activity	Monthly	3 CMI
14	Identified cost per unit of employee's activity/market price per unit of employee's activity	Monthly	2 CMI

Table 5.9 (*Continued*)   Purchasing (Supply Chain) Workers (Clerks, Store Keepers, Helpers, Drivers, Security Guards)

No.	Parameter	Periodicity of Measurement	Suggested Weight and Category
15	Employee's identifiable cost per unit of group's activity/total identifiable cost per unit of group's activity	Monthly	2 CMI
16	Total cost per unit of group's activity/total cost per unit of complete product	Monthly	3 CMI
17	Total identifiable cost per unit of group's activity/total identifiable cost per unit of complete product	Monthly	2 CMI
18	Total cost per unit of group's activity/market price per unit of group's activity	Monthly	2 CMI
19	Employee's allowances and perquisites/total salary	Monthly	2 CMI
20	Retirement benefits contributed by employer/total salary	Monthly	2 CMI
21	Cash salary as percentage of total salary	Monthly	2 EPI
22	Taxable salary as percentage of total salary	Monthly	2 EPI
23	Percentage increase in salary (excluding dearness allowance)/percentage increase in employee's productivity	Annually/six monthly	4 EPI
24	Employee's salary/average salary in region	Six monthly	2 EPI
25	Employee's salary/average salary in industry	Six monthly	2 EPI
26	Employee's salary/average salary in organization	Six monthly	3 EPI
27	Employee's salary/salary of boss	Six monthly	2 EPI
28	Employee's salary/highest salary in organization	Six monthly	2 EPI

(*Continued*)

Table 5.9 (*Continued*) Purchasing (Supply Chain) Workers (Clerks, Store Keepers, Helpers, Drivers, Security Guards)

No.	Parameter	Periodicity of Measurement	Suggested Weight and Category
29	Employee's total cost to company/total cost of hiring contract or temporary worker	Six monthly	3 OMI
30	Employee's ownership rights or shares/average ownership shares per employee of organization	Annually	2 EPI
31	Employee's ownership shares/highest number of shares owned by an employee of organization	Annually	3 EPI
32	Employee's ownership shares/average ownership shares per employee in industry or rival company	Annually	3 EPI
33	Average rate of increase in employee's productivity for past 3 or 5 years/average rate of increase in total salary during past 3 or 5 years	Annually	3 EPI
34	Average rate of increase in employee's total salary during past 3 or 5 years/average rate of increase in total average salary in organization or same class of worker	Annually	3 EPI
35	Average rate of increase in employee's productivity in past 3 or 5 years/average rate of increase in productivity of organization or same class of worker	Annually	4 EPI
36	Employee's idle time salary including holidays as percentage of total salary/average idle time salary as percentage of average total salary	Quarterly	3 EPI
37	Employee's salary per degree of productivity/average salary per degree of productivity of newer workers in same category	Quarterly	3 EPI

Table 5.9 (*Continued*) Purchasing (Supply Chain) Workers (Clerks, Store Keepers, Helpers, Drivers, Security Guards)

No.	Parameter	Periodicity of Measurement	Suggested Weight and Category
38	Average rate of increase in employee's salary during past 6 years/average rate of increase in organization's operating profit over past 6 years	Annually	4 EPI
39	Cost of employee's training and development in areas of skills, attitude, and knowledge	Annually	2 EPI
40	Cost of employee's training and development per unit of his or her extra productivity/average cost of training and development per worker per unit of extra productivity	Annually	3 EPI
41	Employee's annual training and development cost/employee's value addition over past year	Annually	3 EPI
42	Cost of employee's welfare incurred by company/employee's total salary	Annually	2 CMI
43	Employee's take-home salary/total salary	Quarterly	2 EPI
44	Average rate of increase in employee productivity over past 6 years/average rate of increase in cost of infrastructure provided by company over past 6 years	Annually	3 EPI
45	Cost per extra unit produced by employee/average cost per unit of his or her production	Annually	3 CMI
46	Average rate of increase in employee's dearness allowance during past 6 years/average rate of inflation over past 6 years	Annually	2 CMI
47	Average rate of increase in employee savings during past 6 years/average rate of increase in salary during past 6 years	Annually	3 EPI

(*Continued*)

Table 5.9 (*Continued*)   Purchasing (Supply Chain) Workers (Clerks, Store Keepers, Helpers, Drivers, Security Guards)

No.	Parameter	Periodicity of Measurement	Suggested Weight and Category
48	Cost of abnormal waste caused by employee/cost of normal waste incurred by employee	Monthly	2 CMI
49	Percentage use of warehousing facility	Quarterly	3 OMI
50	Percentage use of transport facility	Quarterly	3 OMI
51	Average purchases made per supplier account	Quarterly	3 OMI
52	Average purchase overhead incurred per purchasing employee	Quarterly	2 CMI
53	Average purchase overhead incurred per supplier	Quarterly	2 CMI
54	Average cost saving achieved per supplier, contractor, or vendor account	Quarterly	4 CMI
55	Ratio of purchase worker's total salary to purchase officers' total salary	Quarterly	2 EPI
56	Purchase overhead as percentage of total overhead	Monthly	3 CMI
57	Purchase cost as percentage of total cost of manufacturing	Quarterly	3 CMI
58	Overhead per purchase transaction made by dedicated purchase department versus overhead on direct purchases made by other departments	Quarterly	3 EPI

Table 5.10 General Administration Workers (including Finance, Accounting, and Human Resource Workers)

No.	Parameter	Periodicity of Measurement	Suggested Weight and Category
1	Total cost per unit of activity, process, or operation carried out by employee (total cost = identifiable costs and common costs apportioned to employee's activity)	Weekly	4 CMI
2	Identifiable or direct cost per unit of employee's activity	Weekly	3 CMI
3	Employee's variable (performance-based) wage/total wage	Monthly	4 EPI
4	Overtime wages/total wages	Monthly	2 EPI
5	Idle time wage/total wage	Monthly	2 EPI
6	Dearness allowance/total wages	Monthly	1 CMI
7	Controllable cost per unit of employee's activity, process, or operation	Weekly	3 CMI
8	Cost of value addition per unit of employee's activity, process, or operation	Weekly	4 CMI
9	Value addition cost per unit of employee's activity/total value addition cost per unit of complete product	Weekly	3 OMI
10	Fixed cost per unit of employee's activity	Monthly	2 CMI
11	Employee's total cost per unit of activity/Market price per unit of activity	Monthly	3 CMI
12	Identified cost per unit of employee's activity/market price per unit of activity	Monthly	2 CMI
13	Identifiable cost per unit of employee's group's activity/total identifiable cost per unit of group's activity	Monthly	2 CMI
14	Total cost per unit of group's activity/total cost per unit of complete product	Monthly	3 CMI

(*Continued*)

Table 5.10 (*Continued*)   General Administration Workers (including Finance, Accounting, and Human Resource Workers)

No.	Parameter	Periodicity of Measurement	Suggested Weight and Category
15	Total identifiable cost per unit of group's activity/total identifiable cost per unit of complete product	Monthly	2 CMI
16	Total cost per unit of group's activity/market price per unit of group's activity	Monthly	2 CMI
17	Employee's allowances and perquisites/total salary	Monthly	2 CMI
18	Retirement benefits contributed by employer/total salary	Monthly	2 CMI
19	Cash salary as percentage of total salary	Monthly	2 EPI
20	Taxable salary as percentage of total salary	Monthly	2 EPI
21	Percentage increase in salary (excluding dearness allowance)/percentage increase in productivity	Annually/six monthly	4 EPI
22	Employee's salary/average salary in region	Six monthly	2 EPI
23	Employee's salary/average salary in industry	Six monthly	2 EPI
24	Employee's salary/average salary in organization	Six monthly	3 EPI
25	Employee's salary/salary of boss	Six monthly	2 EPI
26	Employee's salary/highest salary in organization	Six monthly	2 EPI
27	Employee's total cost/total cost of hiring contract or temporary worker	Six monthly	3 OMI
28	Employee's ownership rights or shares/average ownership shares per employee of organization	Annually	2 EPI
29	Employee's ownership shares/highest number of shares owned by employee of organization	Annually	3 EPI

Table 5.10 (*Continued*)   General Administration Workers (including Finance, Accounting, and Human Resource Workers)

No.	Parameter	Periodicity of Measurement	Suggested Weight and Category
30	Employee's ownership shares/average ownership shares per employee in industry or rival company	Annually	3 EPI
31	Average rate of increase in employee's productivity during past 3 or 5 years/average rate of increase in total salary during last 3 or 5 years	Annually	3 EPI
32	Average rate of increase in total salary during past 3 or 5 years/average rate of increase in total average salary in organization or same class of worker	Annually	3 EPI
33	Average rate of increase in employee's productivity during past 3 or 5 years/rate of increase in average productivity of organization or same class of worker	Annually	4 EPI
34	Idle time salary including holidays as percentage of total salary/average idle time salary as percentage of average total salary	Quarterly	3 EPI
35	Employee's salary per degree of productivity/average salary per degree of productivity of newer workers in same category	Quarterly	3 EPI
36	Average rate of salary increases in last 6 years/average rate of increase in operating profit of organization during past 6 years	Annually	4 EPI
37	Cost of employee's training and development in areas of skills, attitude, and knowledge	Annually	2 EPI
38	Cost of employee's training and development per unit of extra productivity/average cost of training and development per worker per unit of average extra productivity	Annually	3 EPI

(*Continued*)

Table 5.10 (*Continued*) General Administration Workers (including Finance, Accounting, and Human Resource Workers)

No.	Parameter	Periodicity of Measurement	Suggested Weight and Category
39	Cost of employee's annual training and development/value addition produced by employee in past year	Annually	3 EPI
40	Cost of welfare incurred for employee/total salary	Annually	2 CMI
41	Take-home salary/total salary	Quarterly	2 EPI
42	Average rate of increase in employee's productivity during past 6 years/average rate of increase in cost of infrastructure over past 6 years	Annually	3 EPI
43	Average rate of increase in employee's dearness allowance during past 6 years/average rate of inflation for past 6 years	Annually	2 CMI
44	Average rate of increase in employee's savings during past 6 years/average rate of increase in salary over past 6 years	Annually	3 EPI
45	Cost of abnormal waste caused by employee/cost of normal waste caused by employee	Monthly	2 CMI
46	General administration overhead/total overhead of organization	Monthly	4 CMI
47	General administration salaries/total salaries of organization	Quarterly	3 EPI
48	General administration costs incurred for every customer department	Quarterly	3 CMI
49	General administration costs for travel activities of every customer department	Quarterly	3 CMI
50	General administration cost for events and meetings for every customer department	Quarterly	3 CMI
51	Percentage of general administration workers in relation to total number of workers in organization	Quarterly	3 EPI

Table 5.10 (*Continued*)   General Administration Workers (including Finance, Accounting, and Human Resource Workers)

No.	Parameter	Periodicity of Measurement	Suggested Weight and Category
52	Percentage of general administration workers in relation to total employees in general administration	Quarterly	2 EPI
53	Annual cost of mechanized work in general administration department/ annual cost of manual work in department	Annually	3 OMI
54	Classification by elements of general administration costs (materials, employee costs, other expenses)	Quarterly	2 CMI
55	General administration cost incurred per employee of organization	Quarterly	3 CMI

Level 2: Knowledge Employees

Table 5.11   Production Leaders

No.	Parameter	Periodicity of Measurement	Suggested Weight and Category
1	Actual cost per unit of activity of employee's group versus budgeted cost	Monthly	4 CMI
2	Actual cost per unit of activity of employee's group versus actual cost for previous day, week, or month	Daily or weekly	3 CMI
3	Actual and budget comparisons for cost elements (materials, remuneration, overhead)	Monthly	3 CMI
4	Actual and budget comparisons for costs of subactivities of employee's group	Monthly	2 CMI
5	Group's actual weighted productivity versus budgeted weighted productivity	Monthly	5 EPI
6	Group's salary cost classification as fixed or variable and productive or unproductive	Monthly	2 CMI
7	Trend of group's productivity growth versus trend of group salary increases	Annually	4 EPI
8	Break-up of employee's variable pay based on his performance, group performance, and company performance	Quarterly	2 EPI
9	Comparison between employee's group and other groups for: (a) cost per unit of value addition; (b) variable pay per head; (c) total pay per head; (d) cost of waste; (e) cost of unproductive time	Quarterly	3 CMI
10	Total cost per unit of activity versus market price per unit of activity	Quarterly	3 CMI
11	Identifiable cost per unit of activity versus market price per unit of activity	Quarterly	3 CMI

Table 5.11 (*Continued*)   Production Leaders

No.	Parameter	Periodicity of Measurement	Suggested Weight and Category
12	Financial value of advantage accrued through suggestions offered by employee's group or improvements achieved by employee's group	Six monthly	3 OMI
13	Cost of knowledge imparted to group versus financial advantage of such knowledge	Annually	3 CMI
14	Comparison of budgeted and actual components of group's salary as: (a) dearness pay; (b) incentive pay; (c) ownership shares; (d) overtime pay; (e) facilitative allowances and perquisites	Quarterly	2 CMI
15	Average rate of increase in group salaries during past 5 years versus average rate of increase in company's profit versus average rate of increase in group's value additions for past 5 years	Annually	4 EPI
16	Identifiable rate of return on the assets used by group (including fixed assets and working capital)	Annually	3 RMI
17	Group's total output at market price/group's total remuneration	Quarterly	3 EPI
18	Incremental or marginal cost per extra unit of group activity versus budgeted cost	Monthly	3 CMI
19	Percentage use of production facility versus budgeted use	Monthly	4 OMI
20	Incremental cost of quality improvement versus incremental profit realized through group quality improvement	Quarterly	2 OMI

(*Continued*)

Table 5.11 (*Continued*) Production Leaders

No.	Parameter	Periodicity of Measurement	Suggested Weight and Category
21	Units produced and supplied by employee's group to next group or sales team	Quarterly	2 OMI
22	Group's inventory carrying cost	Quarterly	3 CMI
23	Average rate of increase in group's productivity during past 5 years versus average rate of growth in overall production productivity of organization	Annually	4 EPI
24	Actual versus targeted cost savings achieved by group	Quarterly	4 CMI
25	Total notional sale value of group's output/total value of assets used	Six monthly	3 OMI

Table 5.12 Marketing and Sales Leaders

No.	Parameter	Periodicity of Measurement	Suggested Weight and Category
1	Actual cost per unit of activity of employee's group versus budgeted cost	Monthly	4 CMI
2	Actual cost per unit of group's activity versus actual cost for previous day, week, or month	Daily or weekly	3 CMI
3	Actual and budget comparisons for cost elements (materials, remuneration, and overhead)	Monthly	3 CMI
4	Actual and budget comparisons for costs of various subactivities of group	Monthly	2 CMI
5	Group's actual weighted versus budgeted weighted productivity	Monthly	5 EPI
6	Group's salary cost break-up as fixed or variable and productive or unproductive	Monthly	2 CMI
7	Trend of group's productivity growth versus trend of group's salary increases	Annually	4 EPI
8	Break-up of employee's variable pay based on his performance, group performance, and company performance	Quarterly	2 EPI
9	Comparison of employee's group and other groups for: (a) cost per unit of value addition; (b) variable pay per head; (c) total pay per head; (d) cost of waste; (e) cost of unproductive time	Quarterly	3 CMI
10	Total cost per unit of activity versus market price per unit of activity	Quarterly	3 CMI
11	Identifiable cost per unit of activity versus market price per unit of activity	Quarterly	3 CMI

(Continued)

Table 5.12 (*Continued*)   Marketing and Sales Leaders

No.	Parameter	Periodicity of Measurement	Suggested Weight and Category
12	Cost of knowledge imparted to group versus financial advantage from such knowledge	Annually	3 CMI
13	Comparison of budgeted and actual components of group salaries as: (a) dearness pay; (b) incentive pay; (c) ownership shares; (d) overtime pay; (e) facilitative allowances and perquisites	Quarterly	2 CMI
14	Average rate of increase in group salaries during past 5 years versus average rate of increase in company profit versus average rate of increase in group's value addition during past 5 years	Annually	4 EPI
15	Identifiable rate of return on assets used by group (fixed assets and working capital)	Annually	3 RMI
16	Group's inventory carrying cost	Quarterly	3 CMI
17	Actual versus targeted cost savings achieved by group	Quarterly	4 CMI
18	Average rate of increase in group productivity in past 5 years versus average rate of increase in company productivity	Annually	4 EPI
19	Total sales/total assets	Quarterly	4 RMI
20	Total sales/average inventory	Quarterly	3 EPI
21	Total sales/average receivables	Quarterly	3 EPI
22	Cost of credit/total sales	Quarterly	3 CMI
23	Average sales per customer account	Quarterly	3 OMI
24	Percentage increase in market share/percentage increase in market	Quarterly	4 EPI

Table 5.12 (*Continued*) Marketing and Sales Leaders

No.	Parameter	Periodicity of Measurement	Suggested Weight and Category
25	Incremental sales/incremental receivables	Quarterly	3 EPI
26	Incremental sales/incremental inventory	Quarterly	3 EPI
27	Sales per new customer accounts/sales per old customer accounts	Quarterly	2 OMI
28	Promotional or brand-based costs/total sales	Quarterly	3 OMI
29	Incremental carrying costs/incremental sales	Quarterly	2 CMI
30	After-sales service costs/total sales	Quarterly	3 CMI
31	Sales of complementary or subsidiary products/sales of main product	Monthly	3 OMI
32	Marketing ROI (gross margin made by marketing employees/total assets)	Quarterly	4 RMI
33	ROI on marketing assets (notional profit made by marketing employees per brand)	Quarterly	3 RMI
34	Gross margin made by marketing employees per brand	Quarterly	3 EPI
35	Marketing costs as percentage of sales at various stages of a product life cycle: (a) development; (b) launch; (c) take-off; (d) consolidation; (e) leadership; (f) revival; and (g) withdrawal	Six monthly or occasionally	3 OMI
36	Incremental brand promotion cost/incremental sales	Quarterly	3 OMI
37	Marketing overhead/production overhead	Quarterly	2 CMI

Table 5.13 Purchasing (Supply Chain) Leaders

No.	Parameter	Periodicity of Measurement	Suggested Weight and Category
1	Actual cost per unit of activity versus budgeted cost of employee's group	Monthly	4 CMI
2	Actual cost per unit of group's activity versus actual cost for previous day, week, or month	Daily or weekly	3 CMI
3	Actual and budget comparisons for cost elements (materials, remuneration, and overhead)	Monthly	3 CMI
4	Actual and budget comparisons for costs of subactivities of employee's group	Monthly	2 CMI
5	Group's actual weighted versus budgeted weighted productivity	Monthly	5 EPI
6	Group's salary cost classification as fixed or variable and productive or unproductive	Monthly	2 CMI
7	Trend of group's productivity growth versus group salary increase trend	Annually	4 EPI
8	Break-up of employee's variable pay based on his or her performance, group performance, and company performance	Quarterly	2 EPI
9	Comparison of employee's group and other groups for: (a) cost per unit of value addition; (b) variable pay per head; (c) total pay per head; (d) cost of waste; (e) cost of unproductive time	Quarterly	3 CMI
10	Total cost per unit of activity versus market price per unit of activity	Quarterly	3 CMI
11	Identifiable cost per unit of activity versus market price per unit of activity	Quarterly	3 CMI

Table 5.13 (*Continued*)   Purchasing (Supply Chain) Leaders

No.	Parameter	Periodicity of Measurement	Suggested Weight and Category
12	Cost of knowledge imparted to employee's group versus financial advantage from such knowledge	Annually	3 CMI
13	Comparison of budgeted and actual components of employee's group salaries as: (a) dearness pay; (b) incentive pay; (c) ownership shares; (d) overtime pay; (e) facilitative allowances and perquisites	Quarterly	2 CMI
14	Average rate of increase in group salaries in past 5 years versus average rate of increase in company's profit versus average rate of increase in group's value addition over last 5 years	Annually	4 EPI
15	Incremental or marginal cost per extra unit of group activity versus budgeted cost	Monthly	3 CMI
16	Group's inventory carrying cost	Quarterly	3 CMI
17	Actual cost savings achieved by group versus targeted savings	Quarterly	4 CMI
18	Total purchases/average credit purchases	Quarterly	3 EPI
19	Credit purchases versus cash purchases	Quarterly	3 OMI
20	Cash discount advantage versus credit advantage	Quarterly	2 EPI
21	Carrying cost versus production loss due to inadequate supply inventory	Quarterly	3 OMI
22	Total purchases/average inventory	Monthly	3 OMI
23	Purchases per supplier account	Monthly	3 OMI
24	Outsourcing versus insourcing costs of items	Quarterly	4 EPI

(*Continued*)

Table 5.13 (*Continued*) Purchasing (Supply Chain) Leaders

No.	Parameter	Periodicity of Measurement	Suggested Weight and Category
25	Idle or unavoidable costs due to outsourcing	Quarterly	3 OMI
26	Percentage of total value addition from vendors versus company's total production cost	Quarterly	4 EPI
27	Vendor productivity versus company productivity	Quarterly	4 EPI
28	Vendor cost per unit versus company cost per unit	Quarterly	3 CMI
29	Vendor ROI versus company ROI	Quarterly	3 RMI
30	Carrying cost per unit versus volume discount from supplier per unit	Quarterly	3 CMI
31	Carrying costs of A items versus B and C items	Quarterly	3 CMI
32	Indirect taxes on purchases/total value of purchases	Quarterly	2 CMI

Table 5.14 General Administration Leaders (including Finance, Accounting, and Human Resources)

No.	Parameter	Periodicity of Measurement	Suggested Weight and Category
1	Actual cost per unit of activity of employee's group versus budgeted cost	Monthly	4 CMI
2	Actual cost per unit of employee's group's activity versus actual cost for previous day, week, or month	Daily or weekly	3 CMI
3	Actual and budget comparisons for cost elements (materials, remuneration, and overhead)	Monthly	3 CMI
4	Actual and budget comparisons for costs of various subactivities of employee's group	Monthly	2 CMI
5	Group's actual weighted versus budgeted weighted productivity	Monthly	5 EPI
6	Group's salary classification as fixed or variable and productive or unproductive	Monthly	2 CMI
7	Trend of group's productivity growth versus trend of group's salary increases	Annually	4 EPI
8	Break-up of employee's variable pay based on his or her performance, group performance, and company performance	Quarterly	2 EPI
9	Comparison of employee's group and other groups of: (a) cost per unit of value addition; (b) variable pay per head; (c) total pay per head; (d) cost of waste; (e) cost of unproductive time	Quarterly	3 CMI
10	Total cost per unit of activity versus market price per unit of activity	Quarterly	3 CMI

(Continued)

Table 5.14 (*Continued*)   General Administration Leaders (including Finance, Accounting, and Human Resources)

No.	Parameter	Periodicity of Measurement	Suggested Weight and Category
11	Identifiable cost per unit of activity versus market price per unit of activity	Quarterly	3 CMI
12	Financial value of advantages from suggestions made or improvements achieved by group	Six monthly	3 OMI
13	Cost of knowledge imparted to group versus financial advantage from such knowledge	Annually	3 CMI
14	Comparison of budgeted and actual components of his group's salary for: (a) dearness pay; (b) incentive pay; (c) ownership shares; (d) overtime pay; (e) facilitative allowances and perquisites	Quarterly	2 CMI
15	Average rate of increase of group salaries during past 5 years versus average rate of increase in company profit versus average rate of increase in group's value additions for past 5 years	Annually	4 EPI
16	Group's total output at market price/group's total remuneration	Quarterly	3 EPI
17	Incremental or marginal cost per extra unit of activity made by the group versus budgeted cost	Monthly	3 CMI
18	Group's actual cost savings versus targeted saving	Quarterly	4 CMI
19	Total notional sale value of group's output/total value of assets used by group	Six monthly	3 OMI
20	General administration overhead/total organization overhead	Monthly	4 CMI

Table 5.14 (*Continued*)   General Administration Leaders (including Finance, Accounting, and Human Resources)

No.	Parameter	Periodicity of Measurement	Suggested Weight and Category
21	General administration salaries/total organization salaries	Quarterly	4 EPI
22	General administration costs for travel activities of all customer departments	Quarterly	2 CMI
23	General administration costs for all customer departments	Quarterly	3 CMI
24	General administration costs for events and meetings organized for all customer departments	Quarterly	3 CMI
25	Number of general administrative workers as percentage of total workers in organization	Quarterly	3 EPI
26	Number of general administrative workers as percentage of total employees in general administration department	Quarterly	3 EPI
27	Annual cost of mechanized work in general administration department/annual cost of manual work in department	Annually	2 OMI
28	General administration cost incurred per employee of organization	Quarterly	2 CMI
29	Leased versus owned assets of general administration department	Six monthly	2 RMI

Level 3: Decision-Making Employees

Table 5.15 Production or Plant Head

No.	Parameter	Periodicity of Measurement	Suggested Weight and Category
1	Actual/budgeted plant productivity	Monthly	4 EPI
2	Actual/budgeted cost per unit for: (a) material; (b) labor; (c) overhead; (d) variable costs; (e) fixed costs; (f) identifiable costs; (g) common costs	Monthly	3 CMI
3	Actual/budgeted cost per activity	Monthly	3 CMI
4	Cost per unit made/cost per unit purchased	Quarterly	3 CMI
5	Total fixed wages versus total variable wages	Quarterly	3 EPI
6	Cost of waste/total cost of production	Monthly	3 CMI
7	Value addition cost/total cost of production	Monthly	3 CMI
8	Total outsourcing cost/total production cost	Quarterly	3 EPI
9	Total production cost/total cost	Monthly	4 CMI
10	Total marketing cost/total production cost	Monthly	3 CMI
11	Annual patent use royalties/annual production cost	Annually	2 OMI
12	Intangible/tangible production assets	Six monthly	2 RMI
13	Fixed factory assets/factory working capital	Six monthly	2 RMI
14	Repair and maintenance cost/total production cost	Quarterly	3 CMI
15	Cost of core operations/cost of support operations	Quarterly	3 CMI
16	Actual cost saving/budgeted cost saving	Quarterly	4 CMI
17	Cost of rejected units/cost of total units produced	Monthly	3 EPI
18	Controllable costs/uncontrollable costs	Quarterly	3 CMI
19	Plant break-even points for: (a) total BEP; (b) operating BEP; (c) cash BEP	Annually	3 OMI
20	Notional ROI of plant	Quarterly	2 RMI

Table 5.16   Sales and Marketing Head

No.	Parameter	Periodicity of Measurement	Suggested Weight and Category
1	Actual/targeted volume	Monthly	5 EPI
2	Actual/targeted sales	Monthly	5 EPI
3	Percent market share increase/percent increase in market	Quarterly	4 OMI
4	Marketing overhead/sales: (a) promotional overhead/sales; (b) sales execution overhead/sales	Monthly	4 CMI
5	Gross profit/sales	Monthly	4 EPI
6	Sales/average receivables	Monthly	3 EPI
7	Sales/average stock	Monthly	3 EPI
8	Sales/marketing assets	Quarterly	3 RMI
9	Sales salaries/sales	Monthly	4 CMI
10	Variable sales salaries/total sales salaries	Monthly	3 EPI
11	Variable sales overhead/fixed sales overhead	Monthly	3 CMI
12	Average actual sales overhead per unit/average budgeted sales overhead per unit	Monthly	3 CMI
13	Brand-based overhead/total sales overhead	Monthly	3 CMI
14	Actual earnings of brand/budgeted earnings of brand	Monthly	4 OMI
15	Capitalized brand-building cost/total sales overhead	Quarterly	3 CMI
16	Overdue receivables/total receivables	Monthly	4 OMI
17	Incremental ratios: (a) extra credit to customers/extra sales; (b) extra promotional costs/extra sales; (c) extra sales/extra stock	Quarterly	3 OMI
18	Marketing ROI: (a) gross profit/total investment; (b) gross profit/marketing assets	Quarterly	4 EPI
19	Total sales/total investment	Quarterly	4 OMI
20	Sales overhead per sales process	Quarterly	3 CMI

Table 5.17 Purchasing (Supply Chain) Head

No.	Parameter	Periodicity of Measurement	Suggested Weight and Category
1	Purchasing overhead/total operating cost	Monthly	4 CMI
2	Total purchases/average credit purchases	Monthly	3 OMI
3	Total purchases/average stock	Monthly	4 OMI
4	Actual/targeted purchasing cost savings	Quarterly	4 CMI
5	Rejected items/total purchases	Monthly	4 OMI
6	Purchasing advantages from: (a) volume discounts; (b) cash discounts; (c) government subsidies; (d) supplier credits; supplier gifts	Quarterly	4 OMI
7	Credit purchases/cash purchases	Monthly	3 OMI
8	Total cost of purchases/total production cost	Quarterly	4 OMI
9	Outsourcing versus insourcing costs	Quarterly	4 OMI
10	Cost savings classified by: (a) vendor; (b) contractor; (c) ancillary; (d) service provider; (e) labor provider	Quarterly	4 CMI
11	Analysis of carrying costs: (a) A items; (b) B items; (c) C items; (d) slow-moving items; (e) dead items	Monthly	3 CMI
12	Recurring and non-recurring vender development costs	Quarterly	3 CMI
13	Costs of purchasing delays	Monthly	3 OMI
14	Notional rate of return of purchasing division	Quarterly	2 OMI
15	Variable salaries of purchasing employees/total salaries of purchasing employees	Quarterly	3 EPI
16	Controllable/uncontrollable costs of purchases	Quarterly	3 CMI
17	Vendor or ancillary's ROI versus company's ROI	Quarterly	3 EPI
18	Total indirect taxes/total purchases	Quarterly	3 CMI
19	Actual capital costs or asset purchases versus budgeted asset purchases	Quarterly	3 RMI
20	Actual versus budgeted cost of every purchase activity	Quarterly	4 CMI

Table 5.18 Finance and Accounting Head

No.	Parameter	Periodicity of Measurement	Suggested Weight and Category
1	Actual versus budgeted weighted average cost of capital	Six monthly	5 EPI
2	Actual versus budgeted earnings from treasury operations	Quarterly	4 EPI
3	Profit or loss from foreign exchange exposure	Quarterly	4 EPI
4	Actual versus budgeted costs for: (a) accounting; (b) auditing; (c) treasury; (d) fund raising; (e) financial monitoring; (f) tax planning; (g) strategic planning	Quarterly	4 CMI
5	Actual versus budgeted direct tax savings	Six monthly	4 EPI
6	Actual versus budgeted reduction in working capital	Quarterly	3 RMI
7	Company capital expenditure savings	Six monthly	3 EPI
8	Profit from sales of fixed assets	Six monthly	3 EPI
9	Overall saving in company's total costs	Quarterly	3 CMI
10	Saving in cost of credit	Quarterly	3 EPI
11	Reductions in receivables	Quarterly	3 EPI
12	Improvement in banking operations through: (a) increases in bank limits; (b) decreases in mortgages offered to banks; (c) reductions in interest payable to banks	Quarterly	4 EPI
13	Variable salary costs for finance and accounting division versus fixed salary costs for division	Quarterly	4 EPI
14	Total variable costs of finance and accounting division versus total fixed costs of division	Quarterly	4 CMI
15	Advantage of computerized control system/cost of system	Quarterly	3 OMI

Table 5.19 Human Resources Managers

No.	Parameter	Measurement Periodicity	Suggested Weight and Category
1	Actual versus budgeted change in total cost of human resources	Quarterly	4 EPI
2	Actual versus budgeted reduction in manpower	Six monthly	4 EPI
3	Actual versus budgeted employee turnover	Six monthly	3 EPI
4	Average cost of training and development per employee	Six monthly	3 CMI
5	Actual versus budgeted days of training per employee	Six monthly	3 OMI
6	Actual versus budgeted cost of Human Resources (HR) activities: (a) Recruitment (b) Training and development (c) Performance appraisal (d) Retention of talent (e) Salary record maintenance (f) Retirement and resignation administration (g) Employee facilities management (h) Employee welfare (i) Industrial relations (j) Government regulation compliance	Six monthly	4 CMI

	(k) Manpower budgeting (l) Retrenchment administration (m) Wage negotiations (n) Productivity and pay benchmarking (o) Innovation in HR management (p) Strategic participation in corporate governance, strategic planning, technology management, HR auditing, HR valuation, ownership sharing, separating ownership from management		
7	Ratio of total employees to HR employees	Annually	4 EPI
8	Total HR division cost/total operating cost of organization	Six monthly	5 CMI
9	Cost of insourcing versus cost of outsourcing HR activities	Six monthly	4 CMI
10	Total cost of training and development versus total quantifiable benefit from training and development	Six monthly	4 CMI

Note: Normally general administration is supervised by the human resources official in most organizations. A general administrator position at the decision-making level is not common. Hence, the Level 2 parameters prescribed for knowledge employees are generally adequate for general administration functions.

Level 4: Entrepreneurial Employees

Top managers should be entrepreneurial because they represent their respective functions individually and represent the organization collectively. Their collective financial performance becomes the performance of the organization. We can measure their collective performance levels based on the parameters in Table 5.20.

Table 5.20 Financial Performance

No.	Parameter	Periodicity Measurement	Suggested Weight and Category
1	Operating profitability (operating profit/sales)	Monthly	5 EPI
2	Investment or capital or asset turnover (sales/total operating investment)	Quarterly	5 RMI
3	Operating ROI (operating profit/operating investment)	Quarterly	5 RMI
4	Owners' or post-tax ROI, Return on shareholder's equity (ROSE), Return on owner's fund (ROOF), Return on equity (ROE), or Return on net worth (RONW) (profit after tax/own funds)	Six monthly	5 RMI
5	Version 1: debt/equity ratio (long-term borrowed funds/own funds)	Six monthly	4 RMI
6	Version 2: debt/equity ratio (long- and short-term borrowed funds/own funds)	Six monthly	4 RMI
7	Debt service coverage ratio (PAT + depreciation + interest)/(interest + loan installment)	Six monthly	4 RMI
8	Solvency or liquidity (current assets/current liabilities)	Monthly	4 RMI
9	Net solvency (current assets excluding dead stock and irrecoverable debts/current liabilities)	Monthly	4 RMI
10	Quick solvency (quickly liquid current assets/quickly payable current liabilities)	Monthly	4 RMI

Table 5.20 (*Continued*) Financial Performance

No.	Parameter	Periodicity Measurement	Suggested Weight and Category
11	Economic value added (EVA) calculated as PAT + interest − cost of capital	Six monthly	4 EPI
12	Total, operating, and cash break-even points (BEPs)	Annually	3 OMI
13	Debtors turnover (sales/average debtors)	Quarterly	3 EPI
14	Stock turnover (sales/average stock price)	Quarterly	3 EPI
15	Working capital turnover (sales/average working capital)	Quarterly	3 RMI
16	Fixed asset turnover (sales/average fixed assets)	Six monthly	3 RMI
17	Operating profit/non-operating profit	Quarterly	3 EPI
18	Operating profit/non-operating ROI	Quarterly	3 EPI
19	Operating profit/non-operating investment	Quarterly	3 RMI
20	Employee productivity calculated as: (a) sales per employee; (b) gross profit per employee; (c) operating profit per employee; (d) PAT per employee	Quarterly	4 EPI
21	Individual product performance based on: (a) sales; (b) contribution; (c) identifiable profit; (d) operating profit; (e) operating ROI; (f) market share; (g) profitability	Quarterly	4 OMI
22	Functional Performance based on: (a) cost per unit of function; (b) functional profitability; (c) functional ROI	Quarterly	3 OMI
23	Incremental Performance: (a) incremental profitability; (b) incremental investment turnover; (c) incremental ROI	Quarterly	4 EPI
24	Contribution/sales	Quarterly	3 OMI
25	Absolute and percentage increases in net worth	Quarterly	4 RMI
26	Intangible assets/tangible assets	Six monthly	3 RMI

(*Continued*)

Table 5.20 (*Continued*) Financial Performance

No.	Parameter	Periodicity Measurement	Suggested Weight and Category
27	Brand performance based on: (a) profitability; (b) ROI; (c) increase in earning power; (d) increase in value	Six monthly	4 OMI
28	Capital market performance based on: (a) earnings per share (EPS); (b) price/earnings ratio (P/E); (c) market capitalization increase; (d) yield ratio; (e) cash earnings per share	Six monthly	3 RMI
29	Product life-cycle costing: (a) R&D; (b) launch; (c) take-off; (d) consolidation; (e) leadership; (f) revitalization; (g) withdrawal	Occasionally and annually	3 OMI
30	Investment appraisal: (a) payback period (PBP); (b) average rate of return (ARR); (c) net present value (NPV); (d) internal rate of return (IRR)	Occasionally and annually	5 RMI
31	Successful capitalization of R&D cost/total R&D cost	Occasionally and annually	3 CMI
32	Top management remuneration/profit before tax	Annually	4 EPI
33	Actual/budgeted volume	Quarterly	5 EPI
34	Change in market share/change in market	Quarterly	4 EPI
35	Organization profit/group profit	Quarterly	3 RMI
36	Organization sales/group sales	Quarterly	3 OMI
37	Organization's ROI versus: (a) group ROI; (b) best ROI in group; (c) lowest ROI in group; (d) best ROI in industry; (e) ROI of closest competitor	Six monthly	3 RMI
38	Entrepreneurial employees/total employees	Annually	4 EPI
39	Remuneration of entrepreneurial employees/total remuneration of all employees	Six monthly	4 EPI
40	Total ownership benefits to employees/total remuneration of all employees	Annually	3 EPI

Table 5.20 (*Continued*)   Financial Performance

No.	Parameter	Periodicity Measurement	Suggested Weight and Category
41	Outsourcing cost/total operating cost	Six monthly	4 OMI
42	Rate of weighted average cost of capital versus average interest rates on loans in market versus lowest rate of interest on loans in market	Six monthly	4 RMI
43	Total committed cost/total operating cost	Quarterly	4 CMI
44	Average rate of income tax paid by organization versus average rate of income tax paid in industry	Six monthly and annually	3 CMI
45	Sensitive or special costs/sales: (a) environmental costs/sales; (b) indirect taxes/sales; (c) promotional costs/sales; (d) value addition costs/ sales; (e) strike or lockout costs/sales; (f) employee cost/sales	Quarterly and occasionally	4 CMI
46	Notional/total costs	Quarterly	3 CMI
47	Foreign exchange exposure losses/total operating costs	Quarterly	4 OMI
48	Dividend paid/PAT	Annually	4 RMI
49	Holding company profit versus subsidiary company profit	Quarterly	3 RMI
50	Profit made through government policies/total profit	Quarterly	3 OMI
51	Losses of profits from: (a) stock-outs; (b) market entries of new competitors; (c) abnormal waste (d) sociopolitical problems; (e) plant breakdowns; (f) supply delivery delays	Annually and occasionally	4 OMI
52	ROI of business managed by owner versus ROI of business managed by manager	Quarterly	4 EPI

Chapter 6

Entrepreneurial Score Card: A Tool for Improving Impacts of Strategic Planning and Execution

Need for Entrepreneurial Score Card (ESC)

Achieving a maximum and sustainable beneficial impact of a strategy execution requires a company to obtain positive answers to the following three questions:

- Is the strategy operationally convenient?
- Is the strategy financially viable?
- Is the strategy useful in the long-run?

The answers may be positive if the selection and execution of a strategy are approached in an entrepreneurial manner. By definition, this approach to selection and execution of a strategy depends on an employee's entrepreneurial performance. In other words, he or she should be competent enough to judge five dimensions of every strategy based on:

- Impact on business growth and associated risks
- Dependence on the available operational and logistic system
- Effect on cost competitiveness

Employee ability to achieve successful execution

Impact on return on investment (ROI) in relation to resources required

An employee who wants to function in an entrepreneurial manner must follow a versatile approach that demands speed, risk-taking capacity, visualization of the end results, networking skills, and commercial acumen. Entrepreneurship is all about sustainable business results and thus must be attained through the entrepreneurial performance management system (EPMS) route.

All junior and senior employees should perform like entrepreneurs so that the collective results of strategy execution equal 100%. A strategic plan is basically derived from the vision and mission of an organization. The plan operates as a comprehensive road map for an entrepreneurial journey to the desired destination defined by the statement of vision. A company that utilizes an ESC system will obtain wide-ranging benefits described below.

Maturity—Strategy execution requires business maturity at all the levels of the organization. That means that even junior employees should have a sufficient commercial background to understand the costs and benefits of their activities.

Balanced performance and growth—An ESC system should facilitate balanced performance and growth of employees that should ultimately lead to balanced growth of the organization. For example, a sales executive should manage ROI on receivables and inventories while focusing on sales volume.

Succession—ESCs should help groom future leaders systematically. A highly effective manager should understand resource utilization in detail. When the manager reaches the top position in a division, he or she will treat it as an independent business, i.e., an investment center.

Collaborative effort—ESC utilization should promote interdivisional or interfunctional collaboration. Eventually collaboration should produce organizational homogeneity and better collective performance. For example, a senior marketing manager should collaborate well with research and development (R&D) and production managers to ensure better and faster product development that will lead to company success.

Organization-wide entrepreneurship—The ESC system is intended to facilitate institutionalization of entrepreneurship across an organization. One benefit is that the company will be able to perform better during difficult times. As an example, factory employees may generate cost-saving suggestions during a recession as a result of thinking entrepreneurially about the overall business model.

Stakeholder benefits—An ESC system should compel all employees at all levels to treat their functions as profit centers. The ultimate benefit to be expected is wealth maximization for employees and all other stakeholders of the organization.

Ten Characteristics of Entrepreneurial Employees

Entrepreneurship is basically a combination of business drive, effective and speedy decision making, risk-taking ability, innovative spirit, and networking skills. In other words, entrepreneurship is all about creating wealth for all stakeholders. The concept of entrepreneurship is presented in Figure 6.1. An entrepreneur displays the following 10 characteristics:

1. Drive for creating wealth
2. Pragmatic vision
3. Ability to make speedy decisions and execute them
4. Risk-taking ability
5. Emphasis on problem solving (action, not discussion)
6. Commercial acumen
7. Networking skills
8. Strategic communication

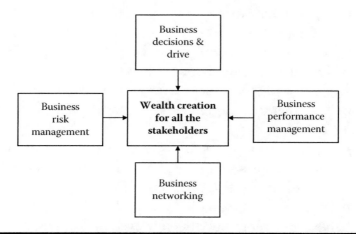

Figure 6.1   Entrepreneurship concept.

9. Behavioral flexibility
10. Rigorous performance measurement and control

As a general rule, an employee who demonstrates at least six of the ten characteristics may be reasonably called entrepreneurial.

Entrepreneurial Behavior Questionnaire

An organization may analyze the behaviors and attitudes of its employees systematically and use the data to determine the degree of entrepreneurship across the organization. The employees may be asked to answer questions like those below to allow management to make a reasonable assessment of the entrepreneurial behavior levels of individual employees and ultimately the level for the entire organization. A "yes" answer to any question indicates an entrepreneurial quality of the respondent.

1. Do you search for an innovative solution to a routine problem?
2. Do you treat uncertainty as a calculated and studied risk?
3. Do you seek a speedy solution to every problem?
4. Do you attempt to create a network and strengthen it for promoting your business?
5. Do you think big?
6. Do you attempt to convert problems into opportunities?
7. Do you perform financial or commercial assessment of every important business decision you undertake?
8. Do you promote your business idea in your organization?
9. Do you look for new business opportunities?
10. Do you try to solve a problem by avoiding systemic hurdles?
11. Do you believe in continuous and impressive growth of business?
12. Do you search for higher benchmarks of performance?
13. Do you believe that knowledge can be converted into a commercial advantage?
14. Do you study the weaknesses and strengths of your competitors?
15. Do you empower your subordinates to make and execute decisions?
16. Do you manage your professional time efficiently?
17. Do you believe in facing competition ruthlessly?
18. Do you believe in simple but effective systems of work?

19. Do you agree that performance-based rewards promote productivity?
20. Do you use interpersonal relations to promote your organization during difficult times?

Classification of Employees Based on Competencies, Roles, and Empowerment Levels

Every organization utilizes four major groups of employees. They all possess and use different competencies and therefore perform various roles requiring several levels of empowerment. The work characteristics of the four groups are illustrated in Figure 6.2.

Systemic or robotic employees include the workers, supervisors, salesmen, junior engineers, and others who follow daily routine instructions and microtargets.

Knowledge employees have more knowledge about their roles and system processes than systemic employees. They raise questions and give constructive feedback for improvisation. They lead small teams and monitor work performance but they do not make decisions. Knowledge employees are usually officers and assistant managers who constitute the junior management of an organization.

Decision-making employees are important because they connect top management to the other levels of employees in an organization. They assist the top executives in preparing strategic plans and converting the plans

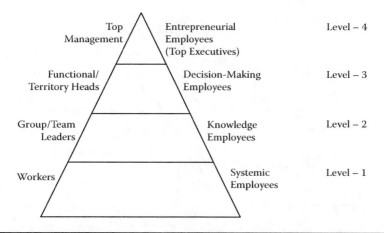

Figure 6.2 Employee types, competencies, and empowerment.

into annual budgets. A decision-making employee is normally considered "number two" in his or her functional or business division. These employees make decisions on behalf of the top management and drive the organization at an operational level. Examples of decision-making employees are senior managers and deputy general managers.

Top executives are the business or functional heads who may serve on a company board. They are responsible for preparing strategic plans and ensuring their execution to reach the desired destination defined by the statement of vision. In other words, they collectively carry out the company mission to materialize their collective vision. See Figure 6.3.

ESCs for All Employees

As noted earlier, the execution of any strategic plan is successful only if all the employees at all the levels of an organization deliver entrepreneurial performance as defined by entrepreneurial score cards (ESCs) for all

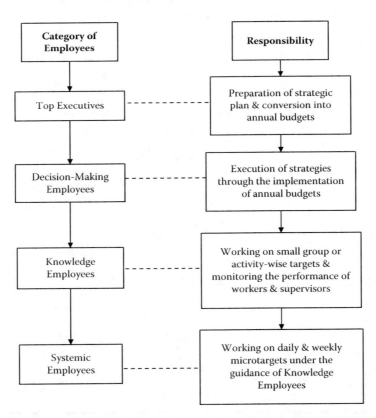

Figure 6.3 Flow of execution of strategic plan.

employees at all levels and in all functions. Before we proceed to ESCs for employees, let us look at a broad ESC that covers an organization (Table 6.1). Recall that an organization must perform in five critical performance areas (discussed in detail in Chapter 2) to ensure that it grows in a balanced manner.

The implementation of strategic plan through an annual budget can be monitored broadly at an organizational level by analyzing the 25 parameters listed in Table 6.1 with and their respective weights. Note that the weights vary from organization to organization based on the criticalities of the parameters. The ultimate execution of annual budgets (whether divisional or functional) is achieved by first analyzing individual performances of employees and converting annual budgets into individual targets as illustrated in Figure 6.4.

ESCs for Four Employee Categories from Six Functions

This section contains ESCs for all the four categories of employees (top executives, decision making, knowledge, and systemic) from six important functions (marketing, production, purchase, finance, human resources, and research and development); see Tables 6.4 through 6.27. The eight principles listed below must be understood.

1. Entrepreneurial performance involves five components:
    Business growth and expansion (core performance)
    Strategic cost management (cost performance)
    Operations, logistics, and technology management (process or activity performance)
    Employee development and satisfaction (human resources-related performance)
    Resource management (fund or investment-related performance)
2. Weights to be attached to the parameters from different areas of performance will vary from level to level and function to function.
3. Actual score against a positive parameter should be calculated using the directly proportionate method, and calculated using the inversely proportionate method in case of a negative parameter (Table 6.2). Total actual score is to be considered for computing variable pay (entrepreneurial reward).
4. A moderately optimistic target (desirable performance) may be called a stretched target. A moderately moderate target (achievable performance) is a normal target an employee must achieve.

Table 6.1 ESC for Organization

No.	Performance Parameter	Moderately Optimistic Target (Desirable) (a)	Moderately Moderate Target (Achievable) (b)	Actual Performance (c)	Weight	Variance 1 (a) − (c)	Variance 2 (b) − (c)	Reasons for Variances and Remedies	Weighted Actual Performance or Actual Score [(c)/(b) × weight]
1	**Business growth and expansion**								
1.1	Volume growth								
1.2	Market share growth								
1.3	Growth through diversification								
1.4	Growth through alteration in value chain (backward, forward, and diagonal integration)								
1.5	Growth through networking (outsourcing, partnering, etc.)								
2	**Strategic cost management**								
2.1	Reduction in cost per unit of product								
2.2	Accuracy and effective use of cost measurement and control systems								
2.3	Use of cost competence as parameter to appraise employees								
2.4	Cost competence as sustained competitive advantage								
2.5	Cost flexibility (discretion to incur cost)								
3	**Employee development and satisfaction**								
3.1	Increase in productivity								
3.2	Increase in satisfaction								

Table 6.1 (*Continued*) ESC for Organization

No.	Performance Parameter	Moderately Optimistic Target (Desirable) (a)	Moderately Moderate Target (Achievable) (b)	Actual Performance (c)	Weight	Variance 1 (a) – (c)	Variance 2 (b) – (c)	Reasons for Variances and Remedies	Weighted Actual Performance or Actual Score [(c)/(b) × weight]
3.3	Productivity as sustained competitive advantage								
3.4	Investment in enhancing productivity (knowledge management, technology improvement, etc.)								
3.5	Productivity initiative as institutional culture								
4	**Operations, logistics, and technology management**								
4.1	Operating profitability								
4.2	Systemic discipline								
4.3	Systemic flexibility								
4.4	Operation improvements								
4.5	Fresh investment in new technology								
5	**Resource management**								
5.1	Operating ROI								
5.2	Owner's ROI								
5.3	Investment turnover								
5.4	Economic value added								
5.5	Liquidity								
							Total Weighted Actual Score		

206 ■ *Strategic Planning, Execution, and Measurement*

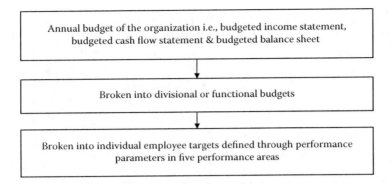

Figure 6.4 Conversion of annual budget into individual targets.

Table 6.2 Score Calculation

Parameter	Moderately Moderate Target	Actual	Weight	Actual Score
Sales volume (positive)	100 units	90 units	30	27 [(90/100) × 30]
Sales expenses/sales (negative)	2.5%	2%	10	12.5 [(2.5/2) × 10]

5. Human resources (HR)-related performance for a senior executive may include his leadership performance also. For a "robotic employee" (factory worker), the HR-related performance shall be based primarily on behavior and self-development efforts. A factory worker need not be assessed for present leadership but may be evaluated for potential leadership.
6. The parameters may be quantitative or qualitative. An effort should be made to use maximum possible quantitative parameters wherever possible to achieve objectivity in evaluation. Qualitative parameters may be converted into quasi-quantitative parameters by using a scale of 0 to 5 where 0 = nil performance and 5 = outstanding performance.
7. On average, two parameters have been suggested for each performance area in the ESC of every employee (ten parameters for each ESC). As we move upward hierarchically from robotic level to knowledge level and further, financial and strategic parameters replace operating parameters.

Table 6.3   Performance Parameter Classifications for All Employee Levels

Employee Level or Category	Performance Parameter Type		
	Strategic	Operational	Financial
Top executives	3 to 4	1 to 3	4 to 5
Decision-making employees	2	4	4
Knowledge employees	Nil	6 to 7	3 to 4
Systemic employees	Nil	7 to 8	2 to 3

A broad classification of parameters for each level of employee is suggested in Table 6.3.

8. Entrepreneurial index is the total score achieved by each employee. The same index may be calculated for functional divisions and for the organization.

Sales and Marketing Employees

Table 6.4 ESC for Robotic or Systemic Employee (Sales Person, Junior Sales Engineer, etc.)

No.	Performance Parameter	Moderately Optimistic Target (Desirable) (a)	Moderately Moderate Target (Achievable) (b)	Actual Performance (c)	Weight	Variance 1 (a) – (c)	Variance 2 (b) – (c)	Reasons for Variances and Remedies	Weighted Actual Performance, i.e., Actual Score [(c)/(b) × weight]
1	**Business growth and expansion**								
1.1	Sales volume								
1.2	Sale price per unit								
2	**Strategic cost management**								
2.1	Sales expenses/sales								
2.2	Incremental cost of gaining new customer								
3	**Employee development and satisfaction**								
3.1	Participation in sales effectiveness training								
3.2	Improvement in sales coordination with colleagues								
4	**Operations, logistics, and technology management**								
4.1	Sales warehouse handling								
4.2	Operational relationships with distributors								
5	**Resource management**								
5.1	Debtors turnover ratio								
5.2	Stock turnover ratio								
							Total Weighted Actual Score		

Table 6.5 ESC for Knowledge Employee (Sales or Marketing Manager, Assistant, or Deputy Marketing Manager)

No.	Performance Parameter	Moderately Optimistic Target (Desirable) (a)	Moderately Moderate Target (Achievable) (b)	Actual Performance (c)	Weight	Variance 1 (a) − (c)	Variance 2 (b) − (c)	Reasons for Variances and Remedies	Weighted Actual Performance, i.e., Actual Score [(c)/(b) × Weight]
1	**Business growth and expansion**								
1.1	Sales volume								
1.2	Sale price per unit								
2	**Strategic cost management**								
2.1	Territory sales expenses/territory sales								
2.2	Carrying cost of inventory and receivables								
3	**Employee development and satisfaction**								
3.1	Composite satisfaction index of subordinates								
3.2	Participation in sales effectiveness training								
4	**Operations, logistics, and technology management**								
4.1	Improvement in sales logistics								
4.2	Improvement in sales reporting by subordinates								
5	**Resource management**								
5.1	Gross margin of territory/sales working capital of territory								
5.2	Gross margin per salesman								
								Total Weighted Actual Score	

Table 6.6 ESC for Decision-Making Employee (Marketing or Senior Marketing Manager, Deputy General Manager of Sales and Marketing)

No.	Performance Parameter	Moderately Optimistic Target (Desirable) (a)	Moderately Moderate Target (Achievable) (b)	Actual Performance (c)	Weight	Variance 1 (a) − (c)	Variance 2 (b) − (c)	Reasons for Variances and Remedies	Weighted Actual Performance, i.e., Actual Score [(c)/(b) × Weight]
1	**Business growth and expansion**								
1.1	Increase in market share								
1.2	Performance of new products (sales volume)								
2	**Strategic cost management**								
2.1	Reduction in promotional cost of each product								
2.2	Reduction in cost of sales logistics								
3	**Employee development and satisfaction**								
3.1	Composite satisfaction index of subordinates								
3.2	Improvement in marketing acumen of territory managers								
3.3	Self-improvement in strategic approach to marketing								
4	**Operations, logistics, and technology management**								
4.1	Improvement in distribution network								
4.2	Improvement in marketing control system								
5	**Resource management**								
5.1	Product-wise ROI								
5.2	ROI on marketing investment								
							Total Weighted Actual Score		

Table 6.7 ESC for Top Executive (General Manager of Sales and Marketing, Marketing Vice President, Marketing Director)

No.	Performance Parameter	Moderately Optimistic Target (Desirable) (a)	Moderately Moderate Target (Achievable) (b)	Actual Performance (c)	Weight	Variance 1 (a) – (c)	Variance 2 (b) – (c)	Reasons for Variances and Remedies	Weighted Actual Performance, i.e., Actual Score [(c)/(b) × Weight]
1	**Business growth and expansion**								
1.1	Increase in total sales turnover								
1.2	Success of new product launches								
1.3	Improvement in brand equity								
2	**Strategic cost management**								
2.1	Sales overhead/sales								
2.2	Promotional cost of new products								
3	**Employee development and satisfaction**								
3.1	Improvement in sales productivity								
3.2	Improvement in marketing leadership skills								
4	**Operations, logistics, and technology management**								
4.1	Improvement in total marketing logistics and coordination with production division								
4.2	Improvement in market intelligence mechanism								
5	**Resource management**								
5.1	Increase in marketing ROI								
5.2	Incremental sales/Incremental sales working capital								
				Total Weighted Actual Score					

Production or Operations Employees

Table 6.8 ESC for Robotic or Systemic Employee (Shop Floor Worker, Supervisor, etc.)

No.	Performance Parameter	Moderately Optimistic Target (Desirable) (a)	Moderately Moderate Target (Achievable) (b)	Actual Performance (c)	Weight	Variance 1 (a) – (c)	Variance 2 (b) – (c)	Reasons for Variances and Remedies	Weighted Actual Performance, i.e., Actual Score [(c)/(b) × Weight]
1	**Business growth and expansion**								
1.1	Output per month or quarter								
1.2	Improvement in quality index								
2	**Strategic cost management**								
2.1	Cost of value addition per unit								
2.2	Waste and scrap reductions								
3	**Employee development and satisfaction**								
3.1	Participation in teamwork								
3.2	Participation in self-development training								
4	**Operations, logistics, and technology management**								
4.1	Attention to process excellence								
4.2	Reduction in delays								
5	**Resource management**								
5.1	Safety of equipment and other assets								
5.2	Output per unit of manufacturing facility or asset								
							Total Weighted Actual Score		

Table 6.9 ESC for Knowledge Employee (Production Supervisor, Team Leader, Assistant Manager)

No.	Performance Parameter	Moderately Optimistic Target (Desirable) (a)	Moderately Moderate Target (Achievable) (b)	Actual Performance (c)	Weight	Variance 1 (a) − (c)	Variance 2 (b) − (c)	Reasons for Variances and Remedies	Weighted Actual Performance, i.e., Actual Score [(c)/(b) × Weight]
1	**Business growth and expansion**								
1.1	Group output per month or quarter								
1.2	Improvement in group average quality index								
2	**Strategic cost management**								
2.1	Average cost of value addition per unit								
2.2	Waste and scrap reduction								
3	**Employee development and satisfaction**								
3.1	Average improvement in employee competence								
3.2	Improvement in leadership impact through training and development								
4	**Operations, logistics, and technology management**								
4.1	Improvement in intragroup coordination								
4.2	Better technological application								
5	**Resource management**								
5.1	Average output per unit of manufacturing facility								
5.2	Output per unit of working capital used								
							Total Weighted Actual Score		

Table 6.10 ESC for Decision-Making Employee (Manager or Deputy Manager in Production, R&M, Logistics Area, etc.)

No.	Performance Parameter	Moderately Optimistic Target (Desirable) (a)	Moderately Moderate Target (Achievable) (b)	Actual Performance (c)	Weight	Variance 1 (a) – (c)	Variance 2 (b) – (c)	Reasons for Variances and Remedies	Weighted Actual Performance, i.e., Actual Score [(c)/(b) × Weight]
1	**Business growth and expansion**								
1.1	Total output for period								
1.2	Output of new product								
1.3	Improvement in output per employee								
2	**Strategic cost management**								
2.1	Average cost per process								
2.2	Average cost per unit of output								
3	**Employee development and satisfaction**								
3.1	Improvement in employee satisfaction								
3.2	Self-improvement in strategic approach to production								
4	**Operations, logistics, and technology management**								
4.1	Improvement in production logistics								
4.2	Process Improvisation leading to commercial benefit								
5	**Resource management**								
5.1	Increase in capacity usage								
5.2	Decrease in factory working capital								
							Total Weighted Actual Score		

Table 6.11 ESC for Top Executive (General Manager, Operations Vice President, Operations Director)

No.	Performance Parameter	Moderately Optimistic Target (Desirable) (a)	Moderately Moderate Target (Achievable) (b)	Actual Performance (c)	Weight	Variance 1 (a) – (c)	Variance 2 (b) – (c)	Reasons for Variances and Remedies	Weighted Actual Performance, i.e., Actual Score [(c)/(b) × Weight]
1	**Business growth and expansion**								
1.1	Total output for period								
1.2	Excellence in supplies to marketing division								
2	**Strategic cost management**								
2.1	Cost reduction per unit of product								
2.2	Sustainable cost reduction through new ventures or experimentation								
3	**Employee development and satisfaction**								
3.1	Improvement in employee retention, talent, and productivity								
3.2	Entrepreneurial leadership of production function								
4	**Operations, logistics, and technology management**								
4.1	Installation or adaptation of new technology								
4.2	Improvement in production logistics through increased output per unit of manufacturing facility								
5	**Resource management**								
5.1	Total production cost/average working capital used in factory								
5.2	Control on capital investment in production function								
								Total Weighted Actual Score	

Purchasing Employees

Table 6.12 ESC for Robotic or Systemic Employee Interacting with Vendors, Contractors, Suppliers, etc.

No.	Performance Parameter	Moderately Optimistic Target (Desirable) (a)	Moderately Moderate Target (Achievable) (b)	Actual Performance (c)	Weight	Variance 1 (a) – (c)	Variance 2 (b) – (c)	Reasons for Variances and Remedies	Weighted Actual Performance, i.e., Actual Score [(c)/(b) × Weight]
1	**Business growth and expansion**								
1.1	Size or volume of purchases handled								
1.2	Number of invoices or suppliers handled								
2	**Strategic cost management**								
2.1	Cost per unit procured								
2.2	Transactional cost per order (ordering cost)								
3	**Employee development and satisfaction**								
3.1	Improvement in overall purchasing productivity								
3.2	Self-improvement through training and development program								
4	**Operations, logistics, and technology management**								
4.1	Time per purchase transaction								
4.2	Improvement in procurement process								
5	**Resource management**								
5.1	Average stock held per period								
5.2	Credits obtained from suppliers								
					Total Weighted Actual Score				

Table 6.13 ESC for Knowledge Employee (Assistant Manager or Group Leader in Purchasing)

No.	Performance Parameter	Moderately Optimistic Target (Desirable) (a)	Moderately Moderate Target (Achievable) (b)	Actual Performance (c)	Weight	Variance 1 (a) – (c)	Variance 2 (b) – (c)	Reasons for Variances and Remedies	Weighted Actual Performance, i.e., Actual Score [(c)/(b) × Weight]
1	**Business growth and expansion**								
1.1	Volume of purchases handled by group								
1.2	Number of new vendors recruited for cost-effective purchases								
2	**Strategic cost management**								
2.1	Cost per unit procured								
2.2	Average cost saving per vendor								
3	**Employee development and satisfaction**								
3.1	Composite satisfaction index of group								
3.2	Self-improvement in leadership skills								
4	**Operations, logistics, and technology management**								
4.1	Improvement in purchase logistics								
4.2	Improvement in transactional accounting and follow-up								
5	**Resource management**								
5.1	Average stock held per period								
5.2	Overall reduction in group's working capital								
							Total Weighted Actual Score		

Table 6.14 ESC for Decision-Making Employee (Senior or Deputy Purchasing Manager)

No.	Performance Parameter	Moderately Optimistic Target (Desirable) (a)	Moderately Moderate Target (Achievable) (b)	Actual Performance (c)	Weight	Variance 1 (a) − (c)	Variance 2 (b) − (c)	Reasons for Variances and Remedies	Weighted Actual Performance, i.e., Actual Score [(c)/(b) × Weight]
1	**Business growth and expansion**								
1.1	Excellence in supplying production division								
1.2	Improvement in vendor performance								
2	**Strategic cost management**								
2.1	Reduction in overall purchase cost								
2.2	Reduction in transactional cost								
3	**Employee development and satisfaction**								
3.1	Increase in employee satisfaction, competence, and productivity								
3.2	Self-improvisation through strategic approach to purchases								
4	**Operations, logistics, and technology management**								
4.1	Improvement in purchasing coordination with production, R&D, and marketing								
4.2	Improvement in procurement process								
5	**Resource management**								
5.1	Total procurement cost/average working capital								
5.2	Advantage of outsourcing over insourcing								
	Total Weighted Actual Score								

Table 6.15 ESC for Top Executive (Director, General Manager or Vice President of Procurement)

No.	Performance Parameter	Moderately Optimistic Target (Desirable) (a)	Moderately Moderate Target (Achievable) (b)	Actual Performance (c)	Weight	Variance 1 (a) – (c)	Variance 2 (b) – (c)	Reasons for Variances and Remedies	Weighted Actual Performance, i.e., Actual Score [(c)/(b) × Weight]
1	**Business growth and expansion**								
1.1	Total procurement volume for period								
1.2	Development of new vendors and ancillaries								
2	**Strategic cost management**								
2.1	Reduction of purchase cost/production cost ratio								
2.2	Reduction of average transactional cost								
3	**Employee development and satisfaction**								
3.1	Improvement in employee entrepreneurial approach								
3.2	Positioning purchasing division as profit center								
4	**Operations, logistics, and technology management**								
4.1	Improvement in overall efficiency of purchasing function								
4.2	Adoption of new purchase monitoring and accounting techniques								
5	**Resource management**								
5.1	Value of purchases/Investment in purchasing division								
5.2	Advantages of outsourcing versus insourcing								
	Total Weighted Actual Score								

Finance and Accounts Employees

Table 6.16 ESC for Robotic or Systemic Employee (Accounts Clerk, Junior Accountant, Clerk, Cashier)

No.	Performance Parameter	Moderately Optimistic Target (Desirable) (a)	Moderately Moderate Target (Achievable) (b)	Actual Performance (c)	Weight	Variance 1 (a) – (c)	Variance 2 (b) – (c)	Reasons for Variances and Remedies	Weighted Actual Performance, i.e., Actual Score [(c)/(b) × Weight]
1	**Business growth and expansion**								
1.1	Number of transactions or invoices handled								
1.2	Improvement in financial or accounting accuracy								
2	**Strategic cost management**								
2.1	Cost per accounting transaction								
2.2	Advantage of insourced accounting activity over outsourced accounting activity								
3	**Employee development and satisfaction**								
3.1	Better internal customer coordination								
3.2	Improvement in functional competency through training and development								
4	**Operations, logistics, and technology management**								
4.1	Improvement in accounting or financial processes								

Table 6.16 (*Continued*) ESC for Robotic or Systemic Employee (Accounts Clerk, Junior Accountant, Clerk, Cashier)

No.	Performance Parameter	Moderately Optimistic Target (Desirable) (a)	Moderately Moderate Target (Achievable) (b)	Actual Performance (c)	Weight	Variance 1 (a) – (c)	Variance 2 (b) – (c)	Reasons for Variances and Remedies	Weighted Actual Performance, i.e., Actual Score [(c)/(b) × Weight]
4.2	Adoption of new accounting technology								
5	**Resource management**								
5.1	Better cash or liquidity management (for cashier) or better accounting support in asset management (for accountant)								
5.2	Improved use of available facilities								
							Total Weighted Actual Score		

Table 6.17 ESC for Knowledge Employee (Assistant Accounts or Audit Manager, Deputy Banking and Finance Manager, etc.)

No.	Performance Parameter	Moderately Optimistic Target (Desirable) (a)	Moderately Moderate Target (Achievable) (b)	Actual Performance (c)	Weight	Variance 1 (a) − (c)	Variance 2 (b) − (c)	Reasons for Variances and Remedies	Weighted Actual Performance, i.e., Actual Score [(c)/(b) × Weight]
1	**Business growth and expansion**								
1.1	Number of transactions handled by group								
1.2	Value added through services to internal customers								
2	**Strategic cost management**								
2.1	Average cost of value added services								
2.2	Reduction in cost per transaction by using better technology or coordination								
3	**Employee development and satisfaction**								
3.1	Composite satisfaction index of group								
3.2	Self-improvement through training and development								
4	**Operations, logistics, and technology management**								
4.1	Improvement in accounting or financial logistics								
4.2	Support to internal customers to improve their accounting or operational control								
5	**Resource management**								
5.1	Better working capital coordination								
5.2	Improved use of available facilities								
							Total Weighted Actual Score		

Table 6.18 ESC for Decision-Making Employee (Senior or Deputy General Manager of Finance, Accounting, Audits, Insurance, Banking, etc.)

No.	Performance Parameter	Moderately Optimistic Target (Desirable) (a)	Moderately Moderate Target (Achievable) (b)	Actual Performance (c)	Weight	Variance 1 (a) – (c)	Variance 2 (b) – (c)	Reasons for Variances and Remedies	Weighted Actual Performance, i.e., Actual Score [(c)/(b) × Weight]
1	**Business growth and expansion**								
1.1	Better accounting and finance deliverables with available human resources								
1.2	Improved quality of accounting, financial, insurance, banking, or auditing products								
2	**Strategic cost management**								
2.1	Reduction in cost of funds or transactional cost								
2.2	Reduction in cost of insurance and other services								
3	**Employee development and satisfaction**								
3.1	Improvement in accounting and financial competence								
3.2	Self-improvement in strategic approach to role								
4	**Operations, logistics, and technology management**								
4.1	Improved application of accounting technology								
4.2	Better coordination with internal customers								

(Continued)

Table 6.18 (*Continued*) ESC for Decision-Making Employee (Senior or Deputy General Manager of Finance, Accounting, Audits, Insurance, Banking, etc.)

No.	Performance Parameter	Moderately Optimistic Target (Desirable) (a)	Moderately Moderate Target (Achievable) (b)	Actual Performance (c)	Weight	Variance 1 (a) – (c)	Variance 2 (b) – (c)	Reasons for Variances and Remedies	Weighted Actual Performance, i.e., Actual Score [(c)/(b) × Weight]
5	**Resource management**								
5.1	Better management of financial resources								
5.2	Innovation in resource mobilization								
							Total Weighted Actual Score		

Table 6.19 ESC for Top Financial Executive (CFO, General Manager or Vice President of Finance, Accounting Control, Banking, Insurance, or Audit & Compliances)

No.	Performance Parameter	Moderately Optimistic Target (Desirable) (a)	Moderately Moderate Target (Achievable) (b)	Actual Performance (c)	Weight	Variance 1 (a) − (c)	Variance 2 (b) − (c)	Reasons for Variances and Remedies	Weighted Actual Performance, i.e., Actual Score [(c)/(b) × Weight]
1	**Business growth and expansion**								
1.1	Suitable delivery of all finance and accounting functions, proportionate to organization's business growth								
1.2	Use of latest financial products								
2	**Strategic cost management**								
2.1	Effective management of funding and insurance costs								
2.2	Cost per transaction								
3	**Employee development and satisfaction**								
3.1	Improved employee competence, output, and satisfaction								
3.2	Strategic participation in corporate leadership								
4	**Operations, logistics, and technology management**								
4.1	Better coordination with internal customers to improve productivity of financial resources								
4.2	Use of advanced technology in accounting control								

(Continued)

Table 6.19 (*Continued*) ESC for Top Financial Executive (CFO, General Manager or Vice President of Finance, Accounting Control, Banking, Insurance, or Audit & Compliances)

No.	Performance Parameter	Moderately Optimistic Target (Desirable) (a)	Moderately Moderate Target (Achievable) (b)	Actual Performance (c)	Weight	Variance 1 (a) − (c)	Variance 2 (b) − (c)	Reasons for Variances and Remedies	Weighted Actual Performance, i.e., Actual Score [(c)/(b) × Weight]
5	**Resource management**								
5.1	Optimal use of available facilities to deliver functional products and performances								
5.2	Better support to internal customers to enhance use of their resources								
							Total Weighted Actual Score		

Human Resource (HR) Employees

Table 6.20 ESC for Robotic or Systemic HR Employee (Clerk, Data Entry Operator, etc.)

No.	Performance Parameter	Moderately Optimistic Target (Desirable) (a)	Moderately Moderate Target (Achievable) (b)	Actual Performance (c)	Weight	Variance 1 (a) – (c)	Variance 2 (b) – (c)	Reasons for Variances and Remedies	Weighted Actual Performance, i.e., Actual Score [(c)/(b) × Weight]
1	**Business growth and expansion**								
1.1	Number of employees served								
1.2	Quality of functional service offered								
2	**Strategic cost management**								
2.1	Cost per transaction of service								
2.2	Contribution to attempts to reduce employee cost								
3	**Employee development and satisfaction**								
3.1	Improved contribution to HR team performance								
3.2	Self-development through training								
4	**Operations, logistics, and technology management**								
4.1	Use of technology in transactional services								
4.2	Efficient handling of coordinative HR services								
5	**Resource management**								
5.1	Increase in advantage of insourced over outsourced HR services								
5.2	Improved performance with available resources								
	Total Weighted Actual Score								

Table 6.21 ESC for Knowledge Employee (Assistant or Deputy HR Manager, etc.)

No.	Performance Parameter	Moderately Optimistic Target (Desirable) (a)	Moderately Moderate Target (Achievable) (b)	Actual Performance (c)	Weight	Variance 1 (a) – (c)	Variance 2 (b) – (c)	Reasons for Variances and Remedies	Weighted Actual Performance, i.e., Actual Score [(c)/(b) × Weight]
1	**Business growth and expansion**								
1.1	Number of employees served by group or team								
1.2	Quality of service offered								
2	**Strategic cost management**								
2.1	Cost of service/total cost of HR division								
2.2	Cost of service/total employee cost								
3	**Employee development and satisfaction**								
3.1	Composite satisfaction index of group or team								
3.2	Self-development for enhancing role as HR professional								
4	**Operations, logistics, and technology management**								
4.1	Improvement in HR performance via process improvisation								
4.2	Better logistic performance through better networking								
5	**Resource management**								
5.1	Quality of value added per dollar spent on HR function								
5.2	Support to internal customers in improving their employee performance through better use of assets								
								Total Weighted Actual Score	

Table 6.22 ESC for Decision-Making Employee (Senior HR Manager or Function Head, e.g., Recruitment, Training and Development, etc.)

No.	Performance Parameter	Moderately Optimistic Target (Desirable) (a)	Moderately Moderate Target (Achievable) (b)	Actual Performance (c)	Weight	Variance 1 (a) – (c)	Variance 2 (b) – (c)	Reasons for Variances and Remedies	Weighted Actual Performance, i.e., Actual Score [(c)/(b) × Weight]
1	**Business growth and expansion**								
1.1	Volume and quality of HR-related services								
1.2	Expansion of role into business partnering with internal customers								
2	**Strategic cost management**								
2.1	Sustainable saving in internal customer HR-related activity cost and employee cost								
2.2	Contribution to cost saving via synergic attempts or structures designed for internal customers								
3	**Employee development and satisfaction**								
3.1	Composite index of subordinate satisfaction								
3.2	Self-development for strategic role in organization								
4	**Operations, logistics, and technology management**								
4.1	Technological application and innovation in HR services								
4.2	Better coordination efforts to improve interfunctional alignments for increased output								

(Continued)

Table 6.22 (*Continued*) ESC for Decision-Making Employee (Senior HR Manager or Function Head, e.g., Recruitment, Training and Development, etc.)

No.	Performance Parameter	Moderately Optimistic Target (Desirable) (a)	Moderately Moderate Target (Achievable) (b)	Actual Performance (c)	Weight	Variance 1 (a) – (c)	Variance 2 (b) – (c)	Reasons for Variances and Remedies	Weighted Actual Performance, i.e., Actual Score [(c)/(b) × Weight]
5	**Resource management**								
5.1	Better value added per dollar spent on HR function								
5.2	Contribution to better asset utilization by internal customers by improving HR effectiveness								
							Total Weighted Actual Score		

Table 6.23 ESC for Top Executive (General Manager, Vice President, or Head of HR Division)

No.	Performance Parameter	Moderately Optimistic Target (Desirable) (a)	Moderately Moderate Target (Achievable) (b)	Actual Performance (c)	Weight	Variance 1 (a) – (c)	Variance 2 (b) – (c)	Reasons for Variances and Remedies	Weighted Actual Performance, i.e., Actual Score [(c)/(b) × Weight]
1	**Business growth and expansion**								
1.1	HR function effectiveness in relation to organizational effectiveness								
1.2	Strategic partnering in long-term HR-related policy framework								
2	**Strategic cost management**								
2.1	HR function cost/total employee cost								
2.2	Reduction in employee costs of internal customers								
3	**Employee development and satisfaction**								
3.1	Composite satisfaction index of all employees								
3.2	Improvement in employee competence across organization								
4	**Operations, logistics, and technology management**								
4.1	Improvisation in HR monitoring of performance management system								
4.2	Technological application and use of innovation to improve HR division performance								

(Continued)

Table 6.23 (*Continued*)   ESC for Top Executive (General Manager, Vice President, or Head of HR Division)

No.	Performance Parameter	Moderately Optimistic Target (Desirable) (a)	Moderately Moderate Target (Achievable) (b)	Actual Performance (c)	Weight	Variance 1 (a) – (c)	Variance 2 (b) – (c)	Reasons for Variances and Remedies	Weighted Actual Performance, i.e., Actual Score [(c)/(b) × Weight]
5	**Resource management**								
5.1	Better organizational resource utilization via improved employee performance								
5.2	Better value added per dollar spent on HR function								
							Total Weighted Actual Score		

Research and Development Employees

Table 6.24 ESC for Robotic or Systemic Employee (Junior Engineer, Researcher, or Research Assistant)

No.	Performance Parameter	Moderately Optimistic Target (Desirable) (a)	Moderately Moderate Target (Achievable) (b)	Actual Performance (c)	Weight	Variance 1 (a) – (c)	Variance 2 (b) – (c)	Reasons for Variances and Remedies	Weighted Actual Performance, i.e., Actual Score [(c)/(b) × Weight]
1	**Business growth and expansion**								
1.1	Completion of milestone of fundamental research								
1.2	Completion of milestone of applied (commercial) research								
2	**Strategic cost management**								
2.1	Cost of items and services purchased to meet milestone								
2.2	Cost of value addition (research progress)								
3	**Employee development and satisfaction**								
3.1	Performance as member of research team								
3.2	Self-development through training and project exposure								
4	**Operations, logistics, and technology management**								
4.1	Contribution to success of group research processes and logistics								
4.2	Improvement in research operations or framework								

(Continued)

Table 6.24 (*Continued*)   ESC for Robotic or Systemic Employee (Junior Engineer, Researcher, or Research Assistant)

No.	Performance Parameter	Moderately Optimistic Target (Desirable) (a)	Moderately Moderate Target (Achievable) (b)	Actual Performance (c)	Weight	Variance 1 (a) – (c)	Variance 2 (b) – (c)	Reasons for Variances and Remedies	Weighted Actual Performance, i.e., Actual Score [(c)/(b) × Weight]
5	**Resource management**								
5.1	Optimal use of research infrastructure								
5.2	Use of working capital for applied research								
							Total Weighted Actual Score		

Table 6.25 ESC for Knowledge Employee (Senior Research Officer, Project Coordinator, Assistant Project Manager)

No.	Performance Parameter	Moderately Optimistic Target (Desirable) (a)	Moderately Moderate Target (Achievable) (b)	Actual Performance (c)	Weight	Variance 1 (a) – (c)	Variance 2 (b) – (c)	Reasons for Variances and Remedies	Weighted Actual Performance, i.e., Actual Score [(c)/(b) × Weight]
1	**Business growth and expansion**								
1.1	Completion of milestone of fundamental research								
1.2	Completion of milestone of applied (commercial) research								
2	**Strategic cost management**								
2.1	Total cost of milestone of research project								
2.2	Cost of value added by research project								
3	**Employee development and satisfaction**								
3.1	Composite satisfaction index of research group								
3.2	Self-development for better research competencies								
4	**Operations, logistics, and technology management**								
4.1	Application of better technology to research								
4.2	Improvement in research logistics to achieve better cooperative efforts								
5	**Resource management**								
5.1	Optimal use of research infrastructure								
5.2	Use of working capital for applied research								
								Total Weighted Actual Score	

Table 6.26 ESC for Decision-Making Employee (Senior or Deputy General Manager or Project Head)

No.	Performance Parameter	Moderately Optimistic Target (Desirable) (a)	Moderately Moderate Target (Achievable) (b)	Actual Performance (c)	Weight	Variance 1 (a) – (c)	Variance 2 (b) – (c)	Reasons for Variances and Remedies	Weighted Actual Performance, i.e., Actual Score [(c)/(b) × Weight]
1	**Business growth and expansion**								
1.1	Approved performance on applied research project								
1.2	Commercial result of project completed in previous period								
2	**Strategic cost management**								
2.1	Total research cost/total sales								
2.2	Total cost of commercial project based on applied research								
3	**Employee development and satisfaction**								
3.1	Employee satisfaction, competence development, and entrepreneurial results								
3.2	Self-development of entrepreneurial approach to research								
4	**Operations, logistics, and technology management**								
4.1	Improvement in research logistics								
4.2	Improvement in collaborative management of joint research projects								

Table 6.26 (*Continued*)   ESC for Decision-Making Employee (Senior or Deputy General Manager or Project Head)

No.	Performance Parameter	Moderately Optimistic Target (Desirable) (a)	Moderately Moderate Target (Achievable) (b)	Actual Performance (c)	Weight	Variance 1 (a) – (c)	Variance 2 (b) – (c)	Reasons for Variances and Remedies	Weighted Actual Performance, i.e., Actual Score [(c)/(b) × Weight]
5	**Resource management**								
5.1	Incremental ROI as a result of value addition from R&D performance								
5.2	Minimization of resource consumption by networking of resource activities								
							Total Weighted Actual Score		

Table 6.27 ESC for Top Executive (Head of R&D Division)

No.	Performance Parameter	Moderately Optimistic Target (Desirable) (a)	Moderately Moderate Target (Achievable) (b)	Actual Performance (c)	Weight	Variance 1 (a) – (c)	Variance 2 (b) – (c)	Reasons for Variances and Remedies	Weighted Actual Performance, i.e., Actual Score [(c)/(b) × Weight]
1	**Business growth and expansion**								
1.1	Entrepreneurial results of current projects and those completed in previous period								
1.2	Incremental value contribution to business from proactive approach to R&D								
2	**Strategic cost management**								
2.1	Research cost/total sales								
2.2	Total fixed cost of R&D division								
3	**Employee development and satisfaction**								
3.1	Improvement in employee research competence								
3.2	Self-development for entrepreneurial approach to research								
4	**Operations, logistics, and technology management**								
4.1	Improvement in research-related operations								
4.2	Application of improved technology to research								

Table 6.27 (*Continued*)   ESC for Top Executive (Head of R&D Division)

No.	Performance Parameter	Moderately Optimistic Target (Desirable) (a)	Moderately Moderate Target (Achievable) (b)	Actual Performance (c)	Weight	Variance 1 (a) − (c)	Variance 2 (b) − (c)	Reasons for Variances and Remedies	Weighted Actual Performance, i.e., Actual Score [(c)/(b) × Weight]
5	**Resource management**								
5.1	Incremental ROI resulting from value addition through R&D performance								
5.2	Combined ROI of research division based on results of projects for current and previous years								
					Total Weighted Actual Score				

Entrepreneurial Rewards

It is obvious that employees will generate entrepreneurial outputs or performances if they receive entrepreneurial rewards such as performance-linked pay plans. Entrepreneurial reward is necessary to achieve the following objectives:

Entrepreneurial drive—Extraordinary performance along with focus on innovation, speed, and flexibility.
Performance recognition—Direct linking of performance and reward so that employees strive continuously to improve output.
Teamwork—Entrepreneurship happens through coordinated efforts. Hence, some fraction of the entrepreneurial reward is to be determined by team performance.
Talent retention—To promote entrepreneurship, employee potential should be recognized through entrepreneurial rewards. Talented employees will remain with an organization if they strongly feel that the organization will provide ongoing entrepreneurial rewards.

Connectivity of individual and organization—An individual should gain proportionately from an organization's prosperity and should also sacrifice for the organization during difficult times.

Entrepreneurial rewards should also depend on organizational dimensions. Three additional factors should be considered in planning entrepreneurial rewards:

1. An employee's value chain, market variables, and criticality of his or her role
2. Product or activity group in which an employee acts as a team member
3. Organization and its business model

The major components of entrepreneurial rewards listed above can be classified into employee, group (or department, division, or location), and organization:

Employee
  Core performance or role
  Support performance or role
  Improvement in competence
  Market conditions

Group, Department, Division, or Location
  Core performance (volume and quality)
  Cost performance
  Support performance
  Improvement in value chain

Organization
  Sales and market share
  Operating profitability
  Investment turnover
  Operating ROI (Return on capital employed [ROCE])
  Owners' ROI
  Debt service coverage
  New product and market development
  Position of organization and positions of products in their life cycles

The weights to be attached to the qualitative and quantitative parameters representing all the three dimensions of entrepreneurial rewards may be prescribed for all the four categories of employees as shown in Table 6.28.

Table 6.28 Weighting of Qualitative and Quantitative Parameters of Entrepreneurial Reward System

Parameter	Total Attached Weights								
	Top Employees		Decision-Making Employees		Knowledge Employees		Systemic Employees	
	Core	Support	Core	Support	Core	Support	Core	Support
Qualitative								
Individual	11	30	15	35	16	37	16	40
Group	10	20	23	20	18	20	9	16
Organization	9	10	7	10	6	8	5	4
Total	30	60	45	65	40	65	30	60
Quantitative								
Individual	19	10	33	10	38	13	48	20
Group	20	18	15	15	15	15	15	14
Organization	31	12	7	10	7	7	7	6
Total	70	40	55	35	60	35	70	40

Knowledge Management for Entrepreneurial Competencies

Strategic planning and execution succeed in an organization when all the employees at all levels and from all functions participate in planning and execution with appropriate maturity. Such maturity is a direct and proportionate result of the initiatives taken by an organization to enhance versatile competencies of its employees through knowledge management processes. Organizations that foster knowledge management know that *to grow is to know*.

Objectives

1. Competence building
2. Interfunction and interlocation reconciliation
3. Integrated and balanced value chain
4. Creation of new opportunities
5. Systemic strength
6. Employee retention

Key Components

Methods of Identifying Knowledge Requirements

Business model alterations
Competence gaps
Management of internal and external risks and uncertainties
New business opportunities
Benchmarking (operational and financial)
Organizational effectiveness (entrepreneurial, social, economical, ethical, and intellectual)

Knowledge Sources

Operational experiences
Competitor performances
Purchased knowledge
Secondary data
Perceptional interactions and interviews
Data from external value drivers (suppliers, contractors, distributors, etc.)
Academic institutions and professional organizations
Customers
Government agencies
Society at large

Knowledge Classification and Storage

Table 6.29 is a knowledge classification aid that illustrates the various types of knowledge and expected application times.

Knowledge Evaluation for Potential Use

It is difficult to predict the long-term use and value of knowledge despite the need to acquire and maintain knowledge as a strategic edge over competitors. The potential acquisition and use of all types of knowledge should be based on the following parameters:

Usefulness of such knowledge at future stages of life cycles of products, organization, and employees
Cost of acquisition of knowledge, cash flow available for acquisition, and capacity to absorb cost over time

Table 6.29   Knowledge Classification Matrix

Factor	Knowledge Use									
	Main Business				Allied Businesses				Future Planning	Sale
	IA	STA	MTA	LTA	IA	STA	MTA	LTA		
Business environment										
Technology										
People processes										
Financial resources										
Competition										
Sociopolitical environment										
Business partners										
Product-related performance										

IA = immediate application.
STA = short-term application.
MTA = medium-term application.
LTA = long-term application.

- Commercial use of knowledge to achieve operational, strategic, and financial advantages
- Competence available and competence required to maintain knowledge, upgrade it, and apply it when required and to pursue new opportunities
- Alternate uses for and commercial value (salability) of knowledge
- Application of intellectual property rights including patent advantages

Knowledge Sharing

Knowledge sharing processes (KSPs) are the most critical components of knowledge management. Their value depends on (1) their technical quality,

Table 6.30 Utilization of Learning Processes

Learning Type	Process	Suitable Use
Entrepreneurial	Case studies and brainstorming	Top and middle management
	Trend analysis	Top and middle management
	Exercises	Junior management
Operational	Cataloguing	Top management
	Interactions	Middle and junior management
	Computer-based training (CBT)	Supervisors
	Group and individual exercises	Supervisors and workers
Attitudinal	Case studies and brainstorming	Top and middle management
	Abstracting	Top and middle management
	Exercises	Junior management
	Workshops	Supervisors and workers

(2) organization culture, (3) individuals' ability and willingness to share knowledge; (4) precise definition of knowledge sharing boundaries; (5) time and cost requirements for sharing; (6) unforeseen events that affect processes and sharing parameters. Table 6.30 summarizes the types of learning, processes, and appropriate use.

Knowledge Application

The phases of commercial application of knowledge are listed in Table 6.31.

Knowledge Sharing Mechanisms

Knowledge management teams (KMTs) should be formed to ensure the transition to and effective use of knowledge sharing processes. The two types are knowledge facilitating teams (Figure 6.5) and knowledge beneficiary teams (Figure 6.6).

Employee Feedback

Obtaining and analyzing employee feedback about knowledge management are critical to the program's continued success. Such feedback should be

Table 6.31 Knowledge Application Phases

No.	Phase
1	Reviewing commercial viability of knowledge acquired by direct involvement of employees using it
2	Possible computer simulations to evaluate process and impact of application
3	Quick feedback from users and possible modification of process
4	Possible application of sample, appraisal of its effectiveness, and implementing modifications for regular use
5	Regular use with continuous monitoring
6	Periodic assessment, feedback, and revision

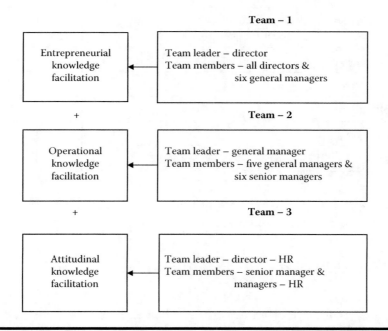

Figure 6.5 Knowledge facilitating teams.

gathered at least twice a year. Employee feedback should cover the following areas:

Awareness of knowledge requirements
Adequacy of knowledge provided
Building ability to use knowledge
Scope for applying knowledge

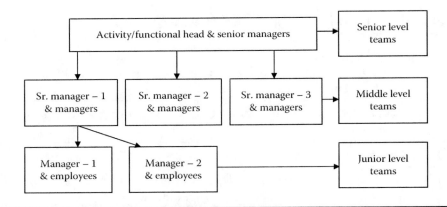

Figure 6.6 Knowledge beneficiary teams.

Table 6.32 Knowledge Management Appraisal Matrix

No.	Knowledge Parameter	Suitability	Acceptability	Impact	Cost	Time	Weight
1	Creation						10
2	Transmission						15
3	Application						10
4	Evaluation						5
	Total						40

  Appraisal of value of knowledge application
  Program rewards and penalties

Management Appraisal

A management audit committee should be appointed to regularly appraise all the implementation factors described above. Table 6.32 is a typical matrix for appraising knowledge management programs.

Promotion of Knowledge Management Efforts

Conformity to knowledge management efforts should be rewarded as a component of the entrepreneurial pay plan. Another type of reward may focus on family support. Exceptional contributions to knowledge management that produce extraordinary business benefits should be rewarded by ownership shares.

Case studies and reports generated through the knowledge management system should be circulated through presentations to various forums and industry publications. Young managers should be encouraged to increase their applications of knowledge through work experience, research, and participation in crisis management. They should also be encouraged to participate in value-adding programs and seminars.

Chapter 7

Comprehensive Case Study

Background and Recent Performance

Asia Cement Limited (ACL) is a listed company based in Mumbai for more than 10 years. It is one of the leaders in the domestic market and known for its product, "Supreme," a medium quality cement brand. Supreme enjoys a 19% market share. ACL also makes lower medium quality cement for the down market and has a 6% market share. During the past 2 years, it started exporting to neighboring countries like Nepal, Sri Lanka, Bangladesh, and Bhutan. ACL has a very powerful vision statement: *Becoming the sole market leader in India in its own segment by achieving the best benchmarks of the cement company.*

ACL is a family-owned company that maintains five factories in different parts of India. Dilip Tolani and his three brothers together own 73% of the shares. A $10 share was quoted at $67 on March 31, 2013. The company is aggressive in its marketing approach but not very professional overall. The Tolanis took little of ACL's profit by way of dividends. Instead they used these internal funds to expand capacity. ACL recently started "feeling the heat" of competition from newcomers in the market who utilize state-of-the-art technology.

Three out of five factories owned by ACL were acquired from failing companies and are very old. ACL does not maintain a proprietary research and development (R&D) facility. The new competitors have joined global players to obtain R&D support. They use new technology and mixtures to frame new products by paying reasonable royalties to the global players.

ACL does not have its own storage facilities and has used leased warehouses for many years. Based on escalations in real estate values, the warehouse owners are demanding substantial rent increases or planning to sell their properties to developers at attractive prices. ACL must seriously analyze its warehousing situation. The Tolani brothers are not known as good pay masters. A number of good managers and officers resigned recently to join competitors at higher salaries.

ACL enjoys a good distribution network. It keeps its distributors happy by paying them a 1% extra margin. However, this strategy may not be very sustainable if ACL's competitors adopt it aggressively. ACL does not do considerable business with customers from the public sector and government departments because it does not believe in offering credit. This has caused considerable opportunity losses over the past 10 years.

ACL will have to expand its production facility by 100% in the next 2 to 3 years to maintain its market share and enjoy the advantages of economy of scale. It may also have to sacrifice a small portion of its present rate of return to accelerate spending for an R&D facility. The conservative approach of financing growth through ploughing profits back into an operation will not support its expansion plan. ACL will have to borrow funds or dilute its present equity to fuel its growth.

ACL's export story is not impressive. It exports low quality cement to underdeveloped neighboring countries. This is not a sustainable business because new competitors are rapidly developing relationships with importers, offering them the benefits of strategic planning and long-term supply chain reliability. The Tolanis are ambitious about their export business and have a vision of growing their exports to other Asian and African countries, but their vision requires work on the company's structure, systems, and strengths. Of course, ample opportunity exists for expanding business in India because India is working very hard to improve its infrastructure and cement is the backbone of infrastructure construction. Tables 7.1 through 7.3 illustrate financial data of the year 2012–13.

Table 7.1 Concise Balance Sheet as of March 31, 2013

Fund Sources	$ (millions)	Investments	$ (millions)
Share capital	60	Fixed assets	100
Reserves	135	Working capital	75
11% bank loan	5	External (non-business investments)	25
Total	200	Total	200

Table 7.2   ACL Production Facilities

Location	Capacity (million tons)	Sales Realization[a] ($ millions)
Thane (old)	1.25	100
Chennai (old)	0.80	80
Bhopal (new)	1.70	170
Ahmadabad (old)	1.50	50
Bhubaneswar (new)	1.00	100
Total	6.25	500

[a] At latest price and 80% capacity use.

Table 7.3   Income Statement for the Year 2012–13

Description	Grade A (Medium Quality Cement) $ (millions)	Grade B (Lower Medium Quality Cement) $ (millions)	Total $ (millions)
Sales	350.00	150.00	500.00
	(2.7 million tons)	(2.3 million tons)	(5 million tons)
Less identifiable operating costs (70% variable)	280.00	135.00	415.00
Identifiable operating profit	70.00	15.00	85.00
Less common operating costs (90% fixed)			20.00
Operating profit (profit before interest & taxes i.e., PBIT)	—	—	65.00
Less interest			0.55
PBT (profit before taxes)	—	—	64.45
Less taxes			19.45
PAT (profit after taxes)	—	—	45.00
Less dividend @ 20% on share capital			12.00
Retained profit	—	—	33.00

Table 7.4 Status Evaluation Summary

Performance Area	Score (0 to 5 Scale)	Average Score in Indian Cement Industry
Business growth and expansion	3.05	3.50
Strategic cost management	3.00	3.60
Operations, logistics, and technology management	2.80	4.00
Employee development and satisfaction	2.70	3.50
Resource management	3.50	3.60
Total score of 25 (% score)	15.05 (60%)	18.20 (73%)

Dilip Tolani and his brothers appointed J&A, a well-known consultant in the field of strategic planning. Tolani realized that a systematic strategic plan was essential to direct the growth of his company. With the help of J&A, the Tolani brothers conducted comprehensive inquiries into all areas of performance of their company and bridged their perceptional gaps about its strengths, weaknesses, opportunities, and growth. Table 7.4 summarizes the status evaluation report prepared by J&A as of January 1, 2013.

It is very obvious from J&A's status evaluation that ACL is performing below average even though its absolute score is 60%. ACL's weak areas are operations, technology, logistics, and employee management. The company does not perform impressively on the fronts of business growth and cost management. Its resource management is satisfactory because it utilizes a conservative approach to capital gearing and could perform at 80% capacity of its factories.

SWOT Analysis

J&A also performed a SWOT (strengths, weaknesses, opportunities, and threats) analysis of ACL using the data revealed by the strategic inquiry exercise. Table 7.5 summarizes the results.

Strategic Plan: Strategies for Five Performance Areas

Dilip Tolani and his brothers got ready to prepare a comprehensive strategic plan for 3 years commencing April 1, 2013. J&A agreed to facilitate and moderate this important exercise. Although J&A recommended a set of optimistic

Table 7.5 Analysis of Strengths, Weaknesses, Opportunities, and Threats

SWOT Factor	Present Status	Future Status
Strengths		
Distribution network	Good	Must be expanded and improved
Control of capital costs	Very good	Should be exploited to finance growth
Reserve position	Very good	Should be used for strategic resource advantage (raising borrowed funds)
Capacity usage	80%	Capacity must be expanded
Unity and homogeneity among promoters	Excellent	Professional approach required
Market share	Good for medium quality product	Should be consolidated further
Foreign market entry	Early effort	Very good growth potential
Spread of production facilities	Well spread throughout country	No facility in north
Product turnover	Reasonable enough to proceed to next phase of consolidation	It is not enough to be an absolute leader; ACL is not in the high product segment; this should be corrected
Weaknesses		
Management quality	Not professional	Should be structured for professional approach
Factories	Three are old	Technological upgrade required
Size of capacity	Insufficient for consolidation	Needs expansion
Use of borrowed funds	Very negligible	Borrowing required to finance expansion
Employee quality and remuneration policy	Below average	Investment in employee education and reward system required

(Continued)

Table 7.5 (*Continued*)   Analysis of Strengths, Weaknesses, Opportunities, and Threats

SWOT Factor	Present Status	Future Status
Use of market opportunities	Not satisfactory	Growth vital to sustain market leadership
Logistics	Warehouses not managed properly	Must build own warehouses to reduce dependence on leased warehouse owners
Product mix	Inadequate for overall presence in market	High quality products must be added to product mix
Exports	Exports only to under-developed countries	Markets in developed countries should also be tapped
Research and development	R&D facilities and efforts are inadequate	A robust R&D facility is required to utilize quality as competitive advantage
Sales to public sector and government departments	None	Public infrastructure demands company product; opportunity cannot be neglected
Organizational structure	Not entrepreneurial at middle and lower levels	Employees should be empowered to take initiatives
Distribution management	Payment of extra margin to distributors may not be sustainable strategy	Strategy may yield good dividends at present but competitors may adopt it; ACL should seek other innovative ways to retain distributors
Opportunities		
1	More exports of better products to obtain higher margins	
2	Expansion of share of domestic market	
3	Good scope for raising borrowed funds required for capacity expansion	
4	Production facility to be created in northern India	
5	Investment in company warehouse should improve quality substantially	

Table 7.5 (*Continued*) Analysis of Strengths, Weaknesses, Opportunities, and Threats

6	Sales to public sector enterprises and government departments should add considerably to market share
7	Entry to high quality product segment should expand domestic market share considerably
8	New distributors may be added for better market penetration in untapped areas
9	Exports to underdeveloped countries can be increased by aggressive pricing
10	International collaborations are possible for R&D initiatives and distribution networks
11	Modernization or replacement of old factories is possible with fresh capital infusion
12	Equity dilution is possible for raising funds through fresh equity at a good premium; present market price is supportive for strategy; ownership dilution degree will not be serious
Threats	
1	Competitors may expand their capacities and increase market shares
2	Logistics management may spin out of control if warehouses are not handled strategically
3	High quality products may become future challenges if not introduced now
4	R&D may provide a competitive edge to competitors
5	Talented employees may leave ACL if not treated well
6	Economies of scale may be lost if present scale of operations is not increased immediately
7	Better technology of competitors may become an unmanageable challenge and result in loss of business
8	Unprofessional style of management may become a big hurdle for developing international joint ventures

and moderate strategies for each objective, Tolani insisted that they initially apply only moderate strategies. The Tolani brothers, J&A consultants, functional heads, and a few brilliant executives formed a strategic planning team (SPT). The executives were selected by J&A in the course of leadership succession planning. Dilip Tolani's son and niece were also involved in the exercise.

The SPT members decided to stay and meet in a hotel for 4 days to conduct their brainstorming sessions and broadly determine the details of goals and strategies for the next 3 years. After this first exercise, the members were divided into groups responsible for the five growth areas (business expansion, cost management, operations management, employee development, and resource management).

Each growth area team included relevant functional representatives. The growth-area teams did a lot of homework on the first draft of the strategic plan and added some valuable ingredients. The entire SPT again met at an outside site for 3 days to finalize the strategic plan. Execution and impact routes were determined for all strategies. The exercise of developing the strategic plan served as an excellent learning experience for new managers and also for the senior Tolanis. J&A played its moderator role very effectively.

The strategic plan covering years 2013–14, 2014–15, and 2015–16 was prepared with a broad forecast of financial measurement of plan impact on performance during the first year. This broad financial forecast was a combination of two key financial statements—income statement and balance sheet. The strategic plan was converted into yearly financial and performance budgets based on function, location, and/or product. This was obviously a comprehensive exercise in budgeting and budgetary control.

Tolani and his colleagues constructed the mission statement: *Expand the capacity to expand the market share, improve technology and logistics to improve delivery, introduce R&D culture to achieve higher quality benchmarks, and develop strong and satisfied human resources.*

The strategic plan covered all five performance areas. It also listed hurdles ACL could expect to face and included a cost–benefit forecast, detailed plan execution schedules, and an exit route. The plan spanning 3 years (April 1, 2013 to March 31, 2016) is summarized in Table 7.6.

Table 7.6 Summary of Strategic Plan

Growth or Performance Strategies	Expected Hurdles	Broad Cost–Benefit Forecast	Execution Schedule	Damage Control or Exit Strategy
Business Growth and Expansion				
Introduce high quality products	Engaging product development competence collaborator at reasonable royalty rate	$5 million investment in R&D center; sales growth over 5 years at 5% per annum	Immediate start; impact to be seen by second year	If start-up fails, collaborate with a university research center
Appoint 50% more new distributors for domestic sales	Resistance from present distributors; readiness of new distributors to make suitable investment and accept challenging conditions	Expected turnover increases of 20, 30, and 40% in 3 years	Start recruiting immediately	Replace existing distributors who resign with new recruits
Increase exports to underdeveloped countries	Stiff competition from new entrants; price cuts expected by customers	Export increases of 10, 15, and 20% expected in 3 years (turnover increases of 2, 3, and 4%)	Effect should be immediate; offer 2% flat discount to importers	Price war may necessitate 3% discount
Start exporting to other Asian and African countries	Competitive price; better quality requirement	Product turnover should increase 10 and 15% in second and third years, respectively	Implement from second year of strategic plan	Pricing may be based on 4 to 5% discount

(Continued)

Table 7.6 (Continued)   Summary of Strategic Plan

Growth or Performance Strategies	Expected Hurdles	Broad Cost–Benefit Forecast	Execution Schedule	Damage Control or Exit Strategy
Collaborate as subcontractor with big global players in developed countries by exporting in their names	Big players may seek large discounts and may reject long-term contracts; government policy may restrict such subcontracts	Present export should increase by 5% with 8 to 10% reduction in sale price	Implement from second year of strategic plan	If global players back out, extra capacity will serve new markets
Start doing business with public sector companies and government departments	Lack of experience of handling such customers; credit recovery	Full exploitation should increase present sales by 10%	Start execution from third year	Delays in credit recovery may affect margins by 10 to 20%

Strategic Cost Management

Develop strategic vendor relationship for 5 years	Sustainable commitment from vendors and projections of future variables may be difficult	Expected saving in cost could be 2% per annum	Develop relationship in 6 months; impact to be seen by second year	Compile list of alternate vendors (at least 20% of present number); be ready to alter relationship parameters
Invest in new technology	Rigidity of policy for funding technological investment by borrowed funds	Capital cost of $10 million; recurring annual saving may be $4 million for 6 years	Execute immediately with immediate benefit by second half of first year	Expected cost saving may be $3 million per annum

Invest in company-owned warehouses	Funding by borrowed funds	Capital cost of $4 million; annual net cost saving should be $1 million for 30 years	Execute immediately for saving to begin from first year	Rentals in some areas may fall substantially, proving ownership to be unviable
Expand present manufacturing capacity	Funding by borrowed funds; lack of professional management	Capital cost should be $50 million for 40% increase in present capacity; annual saving in production cost should be 3%	Start executing immediately; complete work in 1 year; benefit to begin in second year	Delay in execution may cause increase of 5% in capital cost
Improve employee productivity	Worker resistance; cost	Total $1 million cost to be incurred in first and second year; overall benefit projected at $1.5 million per annum for at least 4 years	Start executing immediately; net benefit should be available from third year	Overall impact may fall to 50%

Operations, Logistics, and Technology Management

Invest in new technology, warehouses, and capacity expansion	Discussed in Strategic Cost Management section above			
Improve operational layout for better asset productivity	May be difficult structurally in the old factories Cooperation from the workers may not be optimal	Capital cost of $1 million for old factories; approximate benefit of $1 million per annum for 10 years	Start execution immediately; benefit to accrue in second year	Worker productivity may not match improvement in operational layout

(Continued)

Table 7.6 (Continued) Summary of Strategic Plan

Growth or Performance Strategies	Expected Hurdles	Broad Cost–Benefit Forecast	Execution Schedule	Damage Control or Exit Strategy
Improve distributor's operational efficiency	Old distributors may resist Distributors may not share their operational costs	Approximate cost of $1 million to be incurred in 2 years to increase present turnover by 2% per annum	Execute strategy from first year; benefit after 6 months	Some distributors may continue inefficient practices
Employee Development and Satisfaction				
Introduce professional work culture with improved systems and empowerment throughout organization	Promoters may delay professionalization; workers may be unable to accept culture of improved empowerment	Cost of system improvisation should be $0.2 million to benefit company for 10 years with improved work culture and management	Execute within year; benefit to accrue from second year	Impact may be visible after 2 years; until then some criticism may be expected
Introduce robust system of competence development and knowledge management at all employee levels	Initial responses of promoters and top executives may be negative; work pressure may impede learning environment; investment in learning may not be viewed strategically	Initial capital cost of training facility should be $1 million; recurring annual cost would be $0.2 million for first 3 years; budget must be increased further	Execution in first year should start at top and middle levels simultaneously; training facility should be built within 18 months	Responses from top- and middle-level employees may decrease due to lack of total commitment to learning

Offer performance-based payments to all employees by installing professional performance measurement system	Lower-level workers may resist variability in pay structure; promoters may opt for conservative approach to reasonable budget	Annual increase in total wage cost of 10% should be expected; should result in 20% growth in annual product turnover after second year	Performance measurement system should be developed within 6 months; should be introduced at top- and middle-level employees	Product turnover growth of 20% may not be met; promoters may delay execution
Implement employee welfare programs and foster employee involvement in corporate social responsibility (CSR) programs	May be viewed initially as non-rewarding exercise by promoters and employees; cost of sustaining programs be perceptional challenge	Initially budget of $0.5 million per annum for 2 years, then increase to $1 million per annum; should improve employee satisfaction and company image	Execute strategy from second year after working on other important priorities	Sustaining both the programs depends on promoters' continued interest; employees may treat CSR as formality
Resource Management				
Raise equity by diluting promoters' stakes by 20% and exploit price advantage	Conservative mind-set of promoters may delay approach or reduce dilution to 10%; at time of dilution, price advantage may reduce by 10% due to market response	Approximately $80 million could be raised through this route. But the servicing shall be done only on the shares with a face value of $12 million	Equity for funding expansion should be raised immediately; present market price is beneficial	Issue price may drop as much as 20%; decrease must be factored into plan
Renegotiate with bankers to reduce interest rates from 11 to 9%	Present capital market conditions may not support 2% reduction	Annual saving of $0.10 million should be expected	Start negotiations immediately; benefit should accrue in first quarter of first year	Bank may seek other non-fund-based business from company

(Continued)

Table 7.6 (*Continued*)   Summary of Strategic Plan

Growth or Performance Strategies	Expected Hurdles	Broad Cost–Benefit Forecast	Execution Schedule	Damage Control or Exit Strategy
Assign equity dilution project to present banker's merchant banking division in effort to help interest rate negotiations	Questionable capacity of present banker to get best price for equity dilution	Banker should perform assignment for regular (0.1% of collection) fee and offer reduced interest rate on existing loan	Assign project immediately to bank after confirming its ability to get best price	Banker may perform suboptimally
Create, establish, and use accounting system to capture resource use data for all factories	Present accounting structure and staff may not be able to design and use additional resource control system	One-time cost of developing control system should be around $0.05 million	Establish system within 6 months	Present conservative accounting system may not be modified easily

Measuring Financial Impact of Plan on Company Performance

ACL's strategic plan spans 3 years (2013–2014, 2014–2015 and 2015–2016). The entire plan will have to be divided into three annual budgets (sets of financial targets to be achieved during the year). We shall attempt to measure the financial impact of the strategic plan during the first year. The same methodology may be followed for subsequent years.

Financial measurement requires clear definitions of the major performance and financial targets to be achieved within a year. Table 7.7 sets forth the targets derived from the strategic plan for the first year (2013–2014). Table 7.8 shows a breakdown of operating costs. Table 7.9 details total sales performance. Table 7.10 covers investments in fixed assets. Deferred revenue costs are shown in Table 7.11.

ACL's budgeted financial statements for the first year of the strategic plan (2013–2014) are shown in Tables 7.12 and 7.13. Table 7.12 is an income statement for the year 2013–14. Table 7.13 is a balance sheet as of March 31, 2014.

Table 7.7   First Year Strategic Plan Targets

Performance Area	Target
Sales and Growth	%
Increase in sales by appointing new distributors	20
Increase in sales through more exports to underdeveloped countries	2
Increase in sales through improved operational efficiency of distributors (2% annually starting in second half of year)	1
Incremental Operating Cost	$ Million
Incremental Increases in Operating Costs:	
Amortization of new R&D center cost	1.000
Flat discount of 2% to importers (2% of 2% on present turnover)	0.200
Amortization of new technology cost	1.670
Amortization of capital cost of warehouses	0.130
Cost of improvement in employee productivity	0.500
Cost of improvement in distributor efficiency	0.500
Cost of competence development	0.200
Total Incremental Increase in Operating Cost (a)	4.200

(Continued)

Table 7.7 (Continued)   First Year Strategic Plan Targets

Performance Area	Target
Incremental Decreases in Operating Costs:	
Annual cost saving on account of new technology	2.000
Cost saving on account of own warehouses	1.000
Total Incremental Decrease in Operating Cost (b)	3.000
Net Increase in Operating Cost (a) – (b)	1.200
Financial Cost	**$ Million**
Merchant banker fees (to be written off over 5 years; $0.08 million ÷ 5)	0.016
Savings in interest cost	0.100
Net Decrease in Financial Cost	0.084
Funding Needs	**$ Million**
Total capital expenditure to be incurred during next 3 years	73
Total equity (share capital + premium) to be raised based on approximate face value of $13 million plus premium of $67 million assuming moderate issue price of $60.25 per share ($10 face value plus $50.25 share premium)	80
Dividend Payout	**$ Million**
Dividend at 20% (rate from previous year) on $73 million (old capital of $60 million + new capital of $13 million = $73 million total share capital)	14.6

Table 7.8   Budgeted Total Operating Cost ($ Millions)

	Grade A		Grade B		Total		
Description	V	F	V	F	V	F	Total
Previous year's identifiable cost	196.00	84.00	94.50	40.50	290.50	124.50	415.00
Increase for:							
Volume expansion[a]	45.08	—	21.74	—	66.82	—	66.82
Improvisation and other initiatives[b]	—	0.65	—	0.55	—	1.20	1.20
Inflation at 5%[c]	12.05	4.23	5.81	2.05	17.86	6.28	24.14
Total Identifiable Cost (a)	253.13	88.88	122.05	43.10	375.18	131.98	507.16

Table 7.8 (*Continued*)   Budgeted Total Operating Cost ($ Millions)

Description	Grade A		Grade B		Total		
	V	F	V	F	V	F	Total
Add common operating cost:							
Previous year	—	—	—	—	2.00	18.00	20.00
Increase for:							
Volume expansion	—	—	—	—	0.46	—	0.46
Inflation at 5%[d]	—	—	—	—	0.12	0.90	1.02
Total Common Operating Cost (b)					2.58	18.90	21.48
Total Operating Cost (a + b)					377.76	150.88	528.64

V = Variable. F = Fixed.

[a] Volume expansion (sales growth) is projected at 23% for 2013–2014; product-related variable cost has been increased by 23%.
[b] Cost of improvisation and other initiatives was allocated between Grades A and B in ratio of previous year volumes in million tons.
[c] Average inflation of 5% assumed for 2013–2014.
[d] Increase in identifiable cost due to inflation is calculated by adding 5% of increase due to volume expansion and increase due to improvisation and other initiatives. Same logic is applied to calculating increase in common operating cost due to inflation.

Budgeted Performance and Result Indicators for the Year 2013–2014

- Operating profitability (PBIT/sales) = 14.04%.
- Identifiable profit/sales: Grade A = 20.56%; Grade B = 10.49%.
- Post-tax profitability (PAT/sales) = 9.78%.
- Investment turnover (sales/investment) = 3.51 times. Investment (excluding non-business investment) as of March 31, 2013 was used.
- Operating ROI (PBIT/investment) = 49.35%.
- Owners' ROI (PAT/own fund) = 30.83%. Own fund = share capital + reserves.
- Dividend payout (dividend/PAT) = 24.29%.
- Earnings per share or EPS (PAT/number of shares) = $10.02. Face value of share = $10. Number of shares on March 31, 2011 = 6 million.

- Price/earning ratio (market price of share/EPS) = $5.99.
- Growth in net worth (difference between own fund as of March 31, 2013 and March 31, 2014/own fund as of March 31, 2013) = 64.37%. The substantial growth is based on retained profit and fresh equity raised. If we consider growth *only* from retained profit, it is 23.34% ($45.52 million/$195 million).
- Identifiable operating profit per ton of cement: Grade A ($88.49 million/3.32 million tons) = $26.65; Grade B ($19.35 million/2.83 million tons) = $6.84. Combined Grades A and B ($107.84 million/6.15 million tons) = $17.54.
- Operating profit per ton of cement = $14.04.
- Sale price per ton: Grade A = $129.67; Grade B = $65.19.

Table 7.9 Total Sales Performance for 2013–2014 ($ millions)

Description	Grade A	Grade B	Total
Previous year's sales	350 (2.7 million tons)	150 (2.3 million tons)	500 (5 million tons)
Add: 23% growth expected	80.5 (0.62 million tons)	34.5 (0.53 million tons)	115 (1.15 million tons)
Total Sales, 2013–2014	430.5 (3.32 million tons)	184.5 (2.83 million tons)	615 (6.15 million tons)

Table 7.10 Investment in Fixed Assets for 2013–2014

Fixed Assets	$ (millions)
Tangible	
R&D center	5
Warehouses	4
Manufacturing facility	50
Intangible	
New technology	10
Operational layout	1
Total fixed assets	70

Table 7.11 Investment in Deferred Revenue Cost for 2013–2014

Deferred Revenue Cost	$ (millions)
Cost of improving employee productivity	0.50
Cost of improving operational efficiency of distributors	0.50
Cost of system improvisation	0.20
Cost of new accounting system	0.05
Total Deferred Revenue Cost	1.25

Table 7.12 Budgeted Income Statement for 2013–2014

Description	Grade A	Grade B	Total ($ Millions)
Sales	430.50	184.50	615.00
Less identifiable operating cost	342.01	165.15	507.16
Identifiable operating profit	88.49	19.35	107.84
Less common operating cost	—	—	21.48
Operating profit (PBIT)	—	—	86.36
Less interest and merchant banking fees	—	—	0.47
Profit before tax (PBT)	—	—	85.89
Less tax at 30%	—	—	25.77
Profit after tax (PAT)	—	—	60.12
Less 20% dividend on share capital	—	—	14.60
Retained profit	—	—	45.52

The above indicators must be compared against relevant benchmarks to evaluate the relative performance of ACL. Examples of relevant benchmarks are:

- Average performance in industry
- Best performance in industry
- Performance of close competitor
- Targeted Performance of previous year
- Actual Performance of previous year

Table 7.13 Budgeted Balance Sheet as of March 31, 2014

Sources of Funds		($ Millions)	Investment of Funds		($ Millions)
Share Capital		73.00	Fixed Assets		170.00
Old	60.00		Old	100.00	
New	13.00		New	70.00	
Reserves		247.52	Deferred revenue cost[a]		1.25
Old	135.00				
Share premium on new issue	67.00		Working capital[b]		129.27
Retained profit for 2013–2014	45.52		External investment		25.00
Bank loan (9%)		5.00			
Total		325.52	Total		325.52

[a] Deferred revenue cost is shown as quasi-asset until it is written off.
[b] Amount of working capital has been found as balancing figure.

Execution Route, Time and Resource Mapping, and Impact Route

Every strategy framed in a strategic plan should be supported by a clear and comprehensive execution route (Figure 7.1). The figure depicts a strategy for introducing high quality products and the major milestones along the route.

As discussed earlier, an execution route is a valuable tool for estimating the time and resources required to achieve each milestone of the overall plan execution. It also facilitates the exercise of predicting and handling hurdles and damages at each milestone. One or more alternate routes may be defined if the hurdles and damages along the initial route are unmanageable.

Time and resource mapping (Figure 7.2) is an adjunct to the execution route and helps plan time and cost requirements for execution. An impact route (Figure 7.3) is another tool described in Chapter 3. All three figures focus on the business growth and expansion strategy.

Comprehensive Case Study ■ 269

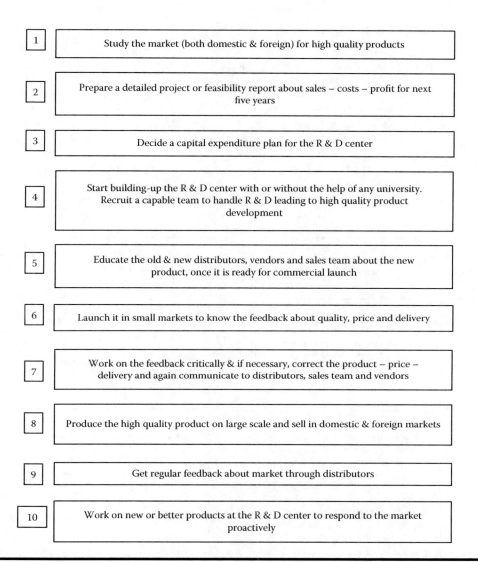

Figure 7.1   Execution route.

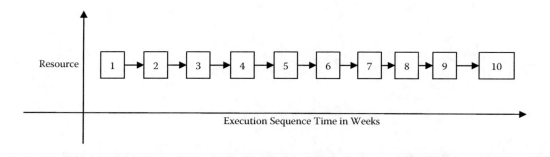

Figure 7.2   Time and resource mapping.

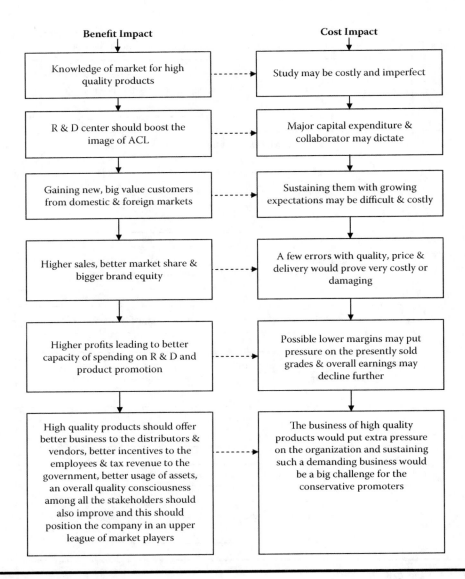

Figure 7.3 Impact route.

Index

A

Academic alliances for technical development, 98
Accounting employees financial monitoring, 183–185, 189
Accounting systems questionnaire for self-inquiry, 43–44
Activity-based costing monitoring tools, 130
Administrative employee financial monitoring, 169–173
Advertising strategy, 87
Annual budgets created from strategic plan projections, 146–149
    individual targets, 206
Asset classification, 100
Asset deployment improvements, 106
Asset utilization monitoring checklist, 140
Automation strategy, 88

B

Backward integration strategy, 84
Banking relationships, 113
Benchmark achievement components, 3
Benchmarking
    connecting operational with financial, 99
    continuous, 5–6
    of earning expectations, 12–14
    goals for higher levels, 98
    improvement in, 5–6
    in organizational culture, 94
    of performance, 30
Benchmarks as goals, 1
BM, see Business model (BM)
Borrowed funds
    for expansion, 115
    utilization, 112
Borrowing cost minimization, 112
Brainstorming for composite performance index, 70
Branding strategy, 91, 97
Budgeted performance in case study, 265–268
Budgets created from strategic plan projections, 146–149
Business development
    components of, 2
    monitoring checklist, 124–126
    questionnaire for self-inquiry, 34ff
Business growth plan in case study, 257–258
Business model (BM)
    behavioral evaluation questionnaire, 37
    development strategies questionnaire, 40
    factors in, 11
    microanalysis, 39
    questionnaire for self-inquiry, 34, 36, 38
    return on investment factor, 12–14
Business plan related to employees skills, 108
Business unit finance involvement, 114

C

Capacity increase strategy, 83
Capital expenditure monitoring checklist, 141–142
Career building methods, 105
    for new employees, 108

Case study
  of Asia Cement Limited, 249ff
  use, 247
Categories of employees, 201–202
CBA (Cost benefit analysis), 75
CMI (Cost management inquiry), 154
Collective impact of strategic areas, 149–150
Company segmentation, 101
Competition barrier strategy, 87
Competitive advantage
  components of, 3
  questionnaire for self-inquiry, 40–44
  sustained by core competence, 15
Competitive strength questionnaire for
    self-inquiry, 23
Competitor challenge strategy, 83
Composite performance index (CPI)
    development, 67–72
Consultant utilization, 109
Coordination team, 80
Core competence
  in strategic plan, 14–15
    to sustain competitive advantage, 15
    upgrading of, 15
Corporate governance self-inquiry
    questionnaire, 22–33
Corporate homogeneity, 8
Cost awareness of employees, 99
Cost center converted to notional profit
    center, 145
Cost competitiveness
  components of, 2
  questionnaire for self-inquiry, 40–44
Cost control strategy, 91
Cost management category, see Strategic
    cost management
Cost management inquiry (CMI)
    category, 154
Cost-benefit analysis (CBA) for strategies, 75
CPI (Composite performance index), 67–72
Crisis management in strategic plan, 75–76

D

Data utilization strategy, 91
Decade-long term for world-class status, 1, 3
  components of, 3

Decision-making employees, 201–202
  ESC for, 210
  ESC for finance and accounts, 223–224
  ESC for human resources, 229–230
  ESC for production, 214
  ESC for purchasing, 218
  ESC for R&D, 236–237
Decision-making employees financial
    monitoring, 186–191
Diagonal integration strategy, 85
Direct marketing development, 97
Distribution channel management
    strategy, 95
Diversification
  reduces overall risk, 15
  strategy, 84

E

Earning expectations, 12–14
Economic value added (EVA) for resource
    management, 117
Economy influence on ROI, 13–14
Economy of scale strategy, 88
Effective systems needed, 4
Eligibility criteria for world-class status, 3
Employee attendance monitoring, 109
Employee awareness
  of funds usage, 115
  of operational costs, 99
  of value chain, 98
Employee career building, 105
Employee development strategy, 105
  encourage top performers, 108
  monitoring checklist, 134–139
  monitoring frequency, 123
  plan in case study, 260–261
  strategies for, 104–111
  studying other organizations, 108
Employee experience limitation strategy, 89
Employee experience mix on teams, 100
Employee family involvement in related
    activities, 108
Employee feedback utilization, 244–245
Employee financial monitoring
  administrative, 169–173
  decision-making employees, 186–191

general administration areas, 183–185
general administration employees, 183–185
knowledge employees, 174–185
marketing and sales areas, 163, 177–179
operating areas, 155–158
production areas, 155–158
production heads, 186
production leaders, 174–176
supply chain, 164–168
supply chain leaders, 180–182
systemic employees, 155–173
Employee involvement in R&D, 102
Employee levels, 152, 154
Employee operational activity groups, 98
Employee participation in quality and cost efforts, 96
Employee productivity inquiry (EPI) category, 154
Employee productivity questionnaire for self-inquiry, 50–59
Employee recognition strategy, 110
Employee salary design, 109
Employee satisfaction monitoring
    checklist, 134–139
    monitoring frequency, 123
Employee satisfaction plan in case study, 260–261
Employee target-setting involvement, 110
Employee training by high-tech operational system, 102
Employee-related issues affecting operations questionnaire for self-inquiry, 47, 50
Empowerment of functional divisions, 145–146
Energy source diversification, 99
Entrepreneurial characteristic, 2
Entrepreneurial drive as objective, 239
Entrepreneurial employees
    characteristics of, 199–200
    financial performance, 192–195
Entrepreneurial performance management system (EPMS), 198
Entrepreneurial reward system
    factors in, 240
    objectives of, 239–240
    structure of, 240–241

Entrepreneurial score card (ESC), 197ff
    for all employees, 202–203
    for decision-making employees in marketing, 210
    for human resources, 227–232
    for knowledge employees in marketing, 209
    for organization, 204–205
    for production employees, 212–215
    for research and development, 233–239
    for robotic employees in marketing, 208
    for systemic employees in marketing, 208
Entrepreneurship strategy for employee involvement, 92
Entrepreneurship within system, 198–200
    behavioral questionnaire, 200–201
Entry barriers strategy, 87
EPI (Employee productivity inquiry), 154
EPMS (Entrepreneurial performance management system), 198
ESC, see Entrepreneurial score card (ESC)
Esteem value, definition, 16
Ethical business practices, 5
EVA (Economic value added), 117
Example setting by management, 102
Execution of strategy
    in case study, 268–270
    financial monitoring of, 145
Execution route in strategic plan, 76
Executive employees, 202
    ESC for finance and accounts, 225–226
    ESC for human resources, 231–232
    ESC for marketing, 210
    ESC for production, 215
    ESC for purchasing, 219
    ESC for R&D, 238–239
Exit strategies in strategic plan, 75–76
Expansion strategies, 83–89

F

Facility sharing strategy, 87
Factoring trade receivables, 115
Finance and accounting heads financial monitoring, 189
Finance and accounts employees ESC, 220–226

Finance employee financial monitoring, 183–185
Financial analysis
  of investment proposals, 114
  of product-wise performance, 26
Financial efficiency factor in performance evaluations, 115
Financial functions outsourced, 114
Financial involvement of business units, 114
Financial management under separate corporations, 115
Financial monitoring; see also specific areas
  administrative, 169–173
  budgets created from strategic plan, 146–149
  collective impact of strategic areas, 149–150
  decision-making employees, 186–191
  entrepreneurial employees, 192–195
  finance and accounting heads, 189
  frequency dependent on type of strategy, 150–152
  general administration employees, 183–185
  human resources managers, 190–191
  knowledge employees, 174–185
  marketing areas, 163
  production heads, 186
  production leaders, 174–176
  supply chain, 164–168
  supply chain heads, 188
  supply chain leaders, 180–182
  system employees, 155–173
Financial performance questionnaire for self-inquiry, 26
Financial projections for impact assessment, 146–149
Flexibility necessary, 4
Flex-time systems, 103
Focus necessary, 3–4
Foreign loan usage, 112
Foreign trade hedging, 113
Forward integration strategy, 85
Franchise operations, 18–19
Functional division empowerment, 145–146
Functional organizational structure development, 97
Funds utilization monitoring checklist, 143

G

General administration financial monitoring, 183–185
General inquiry questionnaire, 22–33
Global leadership, 3
Global preparedness, 3
Government relationship strategy, 90
Growth areas, 2; see also Performance areas
  components of, 2
  for strategies, 73
Growth factors, 66–67
Growth strategies, 83–89
  monitoring frequency, 122

H

Hedging foreign trade, 113
Homogeneity for unity of approach, 8
Horoscope of world-class organization, 3
Housing colony strategy, 103
Human resources employees
  ESC for, 227–232
  financial monitoring of, 190–191
Human resources employees financial monitoring, 183–185
Hurdle management, 75

I

Idle capacity utilization strategy, 86
Impact assessment by financial projections, 146–149
Impact route
  in case study, 268–270
  in strategic plan, 76
Improvisation
  by brainstorming, 79
  in plan strategies, 80
  in process, 5
Information systems questionnaire for self-inquiry, 43–44
Infrastructure sharing strategy, 89, 95
Innovation
  areas of, 1
  components of, 3

culture of, 4
for world-class status, 2
Interest charged on internal funding, 112
Internal funding interest approach, 112
Internal operations as profit centers, 145
Internal rate of return (IRR) for project funding decisions, 116
Intranet implementation, 101
Investment proposal financial analysis, 114
IRR, see Internal rate of return (IRR) for project funding decisions

J

Job rotation questionnaire for self-inquiry, 58

K

KMT, see Knowledge management teams (KMTs)
Knowledge acquisition monitoring checklist, 134
Knowledge application monitoring checklist, 135
Knowledge beneficiary teams, 244–245
Knowledge classification matrix, 242–243
Knowledge employees, 201
    ESC for finance and accounts, 222
    ESC for human resources, 228
    ESC for marketing, 208–211
    ESC for production, 213
    ESC for purchasing, 217
    ESC for R&D, 235
Knowledge employees financial monitoring, 174–185
    marketing and sales leaders, 177–179
    production leaders, 174–176
    supply chain leaders, 180–182
Knowledge facilitating teams, 244–245
Knowledge management
    components of, 242
    monitoring checklist, 136
    objectives of, 241
    rewarding efforts, 246–247
    system utilization, 107
Knowledge management teams (KMTs), 244
Knowledge sharing processes (KSPs), 243–244
Knowledge transfer, see Knowledge management
KSP, see Knowledge sharing processes (KSPs)

L

Labor union negotiations, 105
Leadership styles questionnaire for self-inquiry, 50–52
Learning processes, 243–244
Leasing strategies, 116
Life-cycle costing strategy, 91
Liquidity level maintenance, 113
Loan cost strategy, 93
Logistics management strategies, 94–103
Logistics strategies, 89
    monitoring checklist, 131–133
    plan in case study, 259–260
Longevity for world-class status, 1
Long-term energy keeps vision focus, 7
Lost opportunity cost, 92

M

Management appraisal, 246
Management example-setting, 102
Manager role vs. owner role, 111
Market research strategy, 92
Marketing and sales employee financial monitoring, 177–179
    decision-making employees, 187
Marketing employee financial monitoring, 159–163
Marketing employees, see Knowledge employees
Marketing strategy, 87
    direct channel, 97
Matrix structure development, 96
Mission
    components of, 9–10
    factors in, 11
    steps in, 10–11
    tests of, 9–10
    vision for direction, 9–10

vision for framework, 8
Monitoring of strategy execution
  frequency dependent on type of strategy, 120
  value of, 119–120

N

National economy, 13–14
Networking
  strategy, 86
  for world-class status, 2
New market development strategy, 83
New product strategy, 83
New projects self-funded, 116
Non-operating assets, 100
Notional return on investment, 145

O

OMI, see Operations management inquiry (OMI)
Operating assets, 100
Operational activity groups, 98
Operational advantages questionnaire for self-inquiry, 24
Operational benchmarking, 94
  for compensation, 100
  connecting with performance, 99
Operational catalogs, 94
Operational improvement, 2
Operational system for employee training, 102
Operational team leaders as counselors, 100
Operations control strategy, 94
Operations management
  monitoring checklist, 131–133
  monitoring frequency, 122
  questionnaire for self-inquiry, 45–50
  strategies for, 94–103
Operations management inquiry (OMI), 154
Operations strategic plan in case study, 259–260
Opportunity cost strategy, 92
Optimal strategy questionnaire for self-inquiry, 31
Optimism levels in planning, 72, 74

Organizational loyalty as objective, 240
Organizational structure
  clarity of, 104
  functional development, 97
  for improving logistics, 95
  product-wise, 97
  questionnaire for self-inquiry, 45–46, 59
  simplification of, 109
  strategy, 92
Organizational vision, see Vision
Outsourcing
  of financial functions, 114
  strategy, 85, 88, 94
Overtime culture reduction, 110
Owner role vs. manager role, 111
Owners management of resources, 117

P

Partnership strategy, 85
Perceived value, definition, 16
Performance areas, 2; see also Growth areas
Performance benchmarking, 99
Performance management
  direction for, 7
  questionnaire for self-inquiry, 28–29
Performance monitoring
  questionnaire for self-inquiry, 30
  statement preparation, 120–121
Performance parameter types, 207
Performance planning questionnaire for self-inquiry, 28–29
Performance potential questionnaire for self-inquiry, 34, 35
Performance rating of organization by executives, 67–72
Performance recognition as objective, 239
Performance-based salary strategy, 89
Performance-linked reward system
  questionnaire for self-inquiry, 50–51
Planning of operational elements, 106
Planning skills strategy, 93
Practical value, definition, 16
Product characteristics superiority, 5
Product revitalization strategy, 84
Production employees' ESC, 212–215
Production head financial monitoring, 186

Production in lower cost locations strategy, 89, 96
Production leaders financial monitoring, 174–176
Productivity enhancement
  components of, 2
  empowering managers, 104
  monitoring checklist, 138
  questionnaire for self-inquiry, 55–57, 59
Productivity maintenance efforts, 107
Productivity monitoring checklist, 137
Productivity-based remuneration, 104
Product-wise organizational structure, 97
Profit center concept application, 93
  responsibility accounting, 145
Project funding strategy, 112
Promotion of products strategy, 87
Purchasing employees
  ESC for, 216–219
  financial monitoring of, 164–168, 180–182, 188
Purchasing strategy, 93
  centralization, 101
Pyramid movement, 145

Q

Quantifying parameters and benchmarks, 107
Quarterly reviews, 80–82
Questionnaires for self-inquiry, 22ff

R

Raw material sourcing for quality, 105
Real value, definition, 16
Relationships between strategy components, 6
Relationships within organization questionnaire for self-inquiry, 52, 55
Resource facilitation, 2–3
Resource management
  economic value added (EVA) parameter, 117
  monitoring checklist, 140–143
  monitoring frequency, 123
  by owners, 117
  questionnaire for self-inquiry, 60–66
  strategies for, 112–117
Resource management inquiry (RMI), 154
Resource management plan in case study, 261–262
Resource mapping in case study, 268–270
Retirement benefit development, 109
  salary shifting, 111
Return on investment (ROI) in business model, 12–14
Review of strategic plan, 80–82
Reward system questionnaire for self-inquiry, 50–51
Risky project financed with venture capital, 115
RMI, see Resource management inquiry (RMI)
Robotic employees, see Systemic employees

S

Salary consolidation, 111
Salary plans with specificity to situation, 110
Salary structure design, 109
Sales and marketing head financial monitoring, 187
Sales employees, see Knowledge employees
Sanctity, definition, 5
Securitizing trade receivables, 115
Selectivity necessary, 3–4
Self-funding of new projects, 116
Self-inquiry for strategic planning, 22
Self-management strategy, 94
Sensitivity, definition, 5
Share distribution, 115
Sourcing
  consolidation strategy, 90
  for reduced cost strategy, 93, 95
SPT, see Strategic planning team (SPT)
Standardization strategy, 90
Stock market listing for company, 113
Strategic advantages questionnaire for self-inquiry, 24
Strategic cost management
  in case study, 258–259

monitoring checklist, 127–129
monitoring frequency, 122
need for, 5
questionnaire for self-inquiry, 40–44
strategies for, 92–93
Strategic inquiry into five major areas, 21–22
Strategic plan
   in case study, 252, 256–263
   development process for, 76–79
   lifetime of, 74
   quarterly review, 80–82
   road map function, 73
   roadmap to world-class status, 7
   structure of, 74–76
Strategic planning
   assets for future efforts, 113
   benefits of, 6–7
   collective impact of areas, 149–150
   composite performance index for, 67–72
   independence of areas, 149–150
   knowledge of current state necessary, 21–22
   measuring financial impact in case study, 263–265
   objective of, 145
   questionnaire for self-inquiry, 24, 32
   timeline of, 7
   vision as scope, 7
Strategic planning team (SPT), 76, 77, 79
   preparation of monitoring statements, 120–121
   strategy implementation, 79–80
Strategic position questionnaire for self-inquiry, 25
Strategic thinking exposed to senior level employees, 106
Strategies
   exit routes in event of failure, 6
   relationships between, 6
Strategy execution financial monitoring, 145
Strengths, weaknesses, opportunities, threats analysis, see SWOT (strengths, weaknesses, opportunities, threats) analysis
Structural flexibility necessary, 4

Superiority in product characteristics, 5
Supplier relationship strategy, 88
   suppliers' cost control, 91
Supply chain employee financial monitoring, 164–168
   decision-making employees, 188
   leaders, 180–182
Sustainability necessary, 4
SWOT (strengths, weaknesses, opportunities, threats) analysis
   based on value drivers, 71–72
   in case study, 252, 253–255
Systemic employees, 201
   ESC for finance and accounts, 220–221
   ESC for human resources, 227
   ESC for marketing, 208
   ESC for production, 212
   ESC for purchasing, 216
   ESC for R&D, 233–234
   financial monitoring, 155–173
Systems
   effectiveness of, 4
   major types, 4

T

Talent retention as objective, 239
Target detail communications, 106
Targets as base for strategic planning, 79
Target-setting with employees, 110
Team-building questionnaire for self-inquiry, 52–54
Teamwork as objective, 239
Technical development alliances, 98, 99
Technical organization alliances for technical development, 98
Technology improvement, 105
Technology management, 131–133
   flexibility in, 96
   monitoring checklist, 131–133
   strategies for, 94–103
Technology management plan in case study, 259–260
Threat assessment questionnaire, 34, 36
Time and resource mapping in case study, 268–270
Timeline of strategic planning, 7

U

Union negotiation strategy, 90, 105
    internal union consideration, 107
Unity of approach, 8
Utility value types, 16

V

Value appropriation in strategic role, 16–17
Value chain
    components of, 15–16
    employee education, 98
    franchise as outsourcing of, 18–19
    small enterprises to develop, 99
Value targeting strategy, 87
Value types, 16
Value-added strategy, 92
    evaluating operations, 101
Value-drivers questionnaire for self-inquiry, 27
Value-driving areas, 1
Variable pay questionnaire for self-inquiry, 58
Vendor management strategy, 95
Venture financing for risky projects, 115
Versatile productivity, 104
Visibility necessary for leadership, 7
Vision
    achieved through mission, 9–10
    core competence factor in, 14–15
    departmental statements, 9
    factors that may restrict, 8
    institutionalization of, 8–10
    scope of strategic planning, 7
    at sub-group level, 8
    types of, 9
    vertical nature of, 8–10
Volume capacity increase strategy, 83
Volume growth sustainability, 2

W

Working capital minimization, 114
Working conditions improvement, 105
Working system clarity, 104
World-class organization eligibility criteria, 1
World-class status
    defining, 1
    eligibility criteria for, 3